Memoirs of the American Mathematical Society

Number 434

François Treves

Homotopy formulas in the tangential Cauchy-Riemann complex

Published by the
AMERICAN MATHEMATICAL SOCIETY
Providence, Rhode Island, USA

September 1990 · Volume 87 · Number 434 (second of 3 numbers)

1980 *Mathematics Subject Classification* (1985 *Revision*).
Primary 32, 35.

Library of Congress Cataloging-in-Publication Data

Trèves, François, 1930-
   Homotopy formulas in the tangential Cauchy-Riemann complex/Francois Treves.
   p.   cm. – (Memoirs of the American Mathematical Society, ISSN 0065-9266; no. 434)
   Includes bibliographical references.
   ISBN 0-8218-2496-1
   1. Cauchy-Riemann equations. 2. Homotopy theory. 3. Differential forms. I. Title.
II. Series.
QA3.A57 no. 434
[QA374]
510 s–dc20   90-612
[515'.353]   CIP

Subscriptions and orders for publications of the American Mathematical Society should be addressed to American Mathematical Society, Box 1571, Annex Station, Providence, RI 02901-1571. *All orders must be accompanied by payment.* Other correspondence should be addressed to Box 6248, Providence, RI 02940-6248.

SUBSCRIPTION INFORMATION. The 1990 subscription begins with Number 419 and consists of six mailings, each containing one or more numbers. Subscription prices for 1990 are $252 list, $202 institutional member. A late charge of 10% of the subscription price will be imposed on orders received from nonmembers after January 1 of the subscription year. Subscribers outside the United States and India must pay a postage surcharge of $25; subscribers in India must pay a postage surcharge of $43. Each number may be ordered separately; *please specify number* when ordering an individual number. For prices and titles of recently released numbers, see the New Publications sections of the NOTICES of the American Mathematical Society.

BACK NUMBER INFORMATION. For back issues see the AMS Catalogue of Publications.

MEMOIRS of the American Mathematical Society (ISSN 0065-9266) is published bimonthly (each volume consisting usually of more than one number) by the American Mathematical Society at 201 Charles Street, Providence, Rhode Island 02904-2213. Second Class postage paid at Providence, Rhode Island 02940-6248. Postmaster: Send address changes to Memoirs of the American Mathematical Society, American Mathematical Society, Box 6248, Providence, RI 02940-6248.

10 9 8 7 6 5 4 3 2 1   95 94 93 92 91 90

Contents

Introduction

CHAPTER I
HOMOTOPY FORMULAS WITH EXPONENTIAL
IN THE CAUCHY–RIEMANN COMPLEX

CHAPTER II
HOMOTOPY FORMULAS
IN THE TANGENTIAL CAUCHY–RIEMANN COMPLEX

CHAPTER III
GEOMETRIC CONDITIONS

ABSTRACT

Conjugation of the classical kernels of Bochner–Martinelli & Koppelman– Leray with the FBI transform is used to construct homotopy operators in the tangential Cauchy–Riemann complex. On a real hypersurface in complex space the presence of supporting manifolds is exploited to modify the phase function and ensure the positivity of its imaginary part.

Key Words: *Cauchy–Riemann equations, Homotopy formulas, differential forms*

Work supported in part by NSF Grant DMS–8603171

INTRODUCTION

The purpose of the present work is to show how the FBI (Fourier–Bros–Iagolnitzer) minitransform can be used to derive *local homotopy formulas* for the tangential Cauchy–Riemann differential complex $\overline{\partial}_b$ on certain generic submanifolds of codimension d in $\mathbb{C}^{n+d}$ ($d \geq 1$). By a homotopy formula we mean an identity between differential forms, of the kind

$$(\star) \qquad\qquad f = \overline{\partial}_b Kf + K\overline{\partial}_b f .$$

The solvability problem for the $\overline{\partial}_b$ operator has a long history, going back to the classical papers [Lewy], [K–R], [A–H]. In the late Seventies G. H. Henkin obtained global formulas on the (compact) boundary of a strongly pseudoconvex domain (see [He]). For a description of these, and a generalization to certain weakly pseudoconvex domains, see [Ha–Po]. More recent results, both global and local, are described in [Ai–He], where the reader will find a survey of the field, updated to *circa* 1983, as well as an abundant bibliography. Local results have also been obtained in 1984 by Boggess and Shaw, as well as by Rosay (see [Bo–Sh], [R]). Vanishing theorems for *microlocal* cohomology can be found in [Ka–Sch].

The FBI transform was first introduced in [Br–Ia] to study the microlocal analyticity of distributions. In [B–C–T] it was extended to the study of microlocal *hypo–analyticity*. Actually, in the present work the FBI transform is only used implicitely. What is explicitely used is the conjugate $\mathscr{F}^{-1}\mathscr{K}\mathscr{F}$ of a singular integral operator $\mathscr{K}$ with the FBI minitransform $\mathscr{F}$. The distribution kernel associated with the operator $\mathscr{K}$ is a modification of the classical Bochner–Martinelli–Koppelman–Leray kernel (see [He–L]). The FBI minitransform $\mathscr{F}$ is a simplified version of the transform introduced in [B–C–T] , especially adapted to the present situation: in the base it only involves integration along d–dimensional submanifolds (rather than $(n+d)$–dimensional ones). It also involves an integration along the fibres of the *characteristic set* of the structure, which is a real vector bundle over Σ , of rank d.

In the important hypersurface case, defined by the fact that $d = 1$, it is a line bundle, and one may integrate separately in the half bundle $\sigma > 0$ or in the half bundle $\sigma < 0$. M. S. Baouendi and the author noticed, not long ago, that the FBI minitransform could be used to advantage to study the holomorphic extension of CR functions. The subject of the present study is the natural follow–up of that observation: when $d = 1$ a one–sided homotopy formula will imply the extension, to one or the other side of the hypersurface Σ, of CR cohomology classes (on all this see [T]).

The article is organized as follows:

In Ch. I we restrict our attention to the Cauchy–Riemann operator $\bar{\partial}$ in a convex domain $\mathscr{D}$ in $\mathbb{C}^n$ (where the coordinates are $z_1, ..., z_n$). We establish formulas of the Bochner–Martinelli–Koppelman–Leray (or Henkin–Ramirez) kind, after insertion of an exponential weight–function. The exponent is not necessarily holomorphic with respect to any of the variables; we allow its complex Hessian to change sign. Thus our formulas can be looked upon as a generalization of the formulas in [Be–And]. In view of the applications in Ch. II total precision is required, not only in the expressions of the kernels but also in those of the "error terms". These formulas give information on the growth at infinity of the solutions to the Cauchy–Riemann equations (in spaces of differential forms). It is not excluded that applications to pure complex variable theory (as opposed to CR theory) will eventually be found.

In Ch. II we turn our attention to CR structures. This calls into play d additional complex variables w_k and the real variables $s_k = \mathscr{R}e\, w_k$ ($k = 1,...,d$). The formulas of Ch. I are routinely extended and conjugated with the FBI minitransform, as indicated earlier. In the FBI transform the integration is carried out in s–space $\mathbb{R}^d$ and in the dual space, where the variables is denoted by $\sigma = (\sigma_1, ..., \sigma_d)$ (in half the dual space, $\mathbb{R}_+$ or $\mathbb{R}_-$, when $d = 1$). The construction culminates in the results gathered in Sect. II.7: they assert the existence, and provide the expression, of homotopy operators or, in certain cases, of right–inverses for the tangential Cauchy–Riemann operator $\bar{\partial}_b$ under a set of technical conditions (bearing on the "phase functions" of the FBI integral operators). In passing we must mention that we get rid of certain error terms originating with the s–part of the boundary of the domain of integration by means of a "pinching technique" introduced in [B–T] to study unique continuation of CR functions (see Sections II.4, II.5). This method imposes an additional condition (see (II.4.1)) on the submanifold Σ , condition which is always satisfied when Σ is a hypersurface or when Σ is real analytic but which otherwise might not be, as shown in [J].

The technical hypotheses, in the theorems of Sect. II.7, are not transparent. Ch. III is devoted to showing how they follow from relatively simple geometric properties. The first three sections of Ch. III are entirely devoted to the hypersurface case. In Sect. III.1 we show that the version of the central hypothesis used in practice, which requires that the imaginary part of the phase–function, in the FBI integral operators, be ≥ 0 , is invariant under local biholomorphism. The central hypothesis is equivalent, loosely speaking, to the existence of support functions. Thus we are unable to handle weakly pseudoconvex structures of the kind considered in [K– N]. In Sect. III.2 we define supporting, and strictly supporting, manifolds of holomorphic type ν . The existence of such manifolds represents a kind of convexity, or of concavity, in ν complex directions. When, at the central point, the Levi form of the hypersurface Σ has ν_+ eigenvalues > 0 and ν_- eigenvalues < 0 , then Σ admits strictly supporting manifolds from below of holomorphic type ν_+ and from above, of holomorphic type ν_- (for a suitable definition of "above" and "below"). But, as the reader might well imagine, strictly supporting manifolds exist also in cases where the Levi form does not have the requisite number of eigenvalues of one sign or the other, or even when the Levi form at the central point vanishes.

In Section III.3 we hypothesize that the hypersurface $\Sigma \subset \mathbb{C}^{n+1}$ admits sup–porting manifolds from below, of holomorphic type ν_+ . We look at differential forms of type $(n+1,q)$ (rather than $(0,q)$ for reasons of convenience in the notation: on $(n+1,q)$–forms the operator $\overline{\partial}_b$ can be equated to the exterior differential). We show that there is a one–sided homotopy formula when $q > n-\nu_+$ (Theorems III.3.1,

III.3.2). Next we hypothesize that Σ admits strictly supporting manifolds from above, of holomorphic type ν_- . In this case the one–sided homotopy formula is valid provided $q < \nu_- - 1$; when $q = \nu_- - 1$ there is a microlocal right–inverse of $\overline{\partial}_b$ acting on closed forms (on the appropriate side of the characteristic bundle; Th. III.3.3). These microlocal results entail the obvious local formulas. We note in passing that if Σ is a strongly pseudoconcave hypersurface in $\mathbb{C}^3$, $i.\ e.$, $n = \nu_- = 2$, and if $q = 1$, we do not get a homotopy formula but we do get a right–inverse for $\overline{\partial}_b$. This is consistent with the example in [Na–R] .

Sect. III.4 looks at a generic manifold Σ of codimension $d \geq 2$. We hypothesize that the Levi form of Σ has exactly 2ν nonzero eigenvalues (and therefore ν positive, ν negative and $n{-}2\nu$ null eigenvalues) at every characteristic covector $at\ the\ origin$. We show that the local homotopy formula is valid if $q \notin [\nu{-}1, n{-}\nu]$, whereas there is a right–inverse to $\overline{\partial}_b$ if $q = \nu - 1$. When the codimension of the manifold Σ is ≥ 2 we do not make any attempt at microlocalization, which explains why we need a hypothesis on the Levi form that involves the whole fibre at the origin of the charateristic set.

I would like to thank my students Gabor Francsics and Johannes A. Petersen for pointing out to me numerous mistakes in the successive revisions of this work.

CHAPTER I
HOMOTOPY FORMULAS WITH EXPONENTIAL
IN THE CAUCHY–RIEMANN COMPLEX

1

I.1. THE CAUCHY–RIEMANN COMPLEX IN $\mathbb{C}^n$. NOTATION

Throughout this work $z = (z_1, ..., z_n)$ shall denote the variable point in $\mathbb{C}^n$ ($n \geq 1$). We write $x_i = \mathscr{R}e\, z_i$, $y_i = \mathscr{I}m\, z_i$, $x = (x_1, ..., x_n)$, $y = (y_1, ..., y_n)$, $\bar{z} = x - \imath y$, etc.; $\mathscr{D}$ shall be an open subset of $\mathbb{C}^n$.

We shall deal with differential forms, or with currents, in $\mathscr{D}$, of type (or bidegree) (p,q) :

$$(\text{I.1.1}) \qquad f = \sum_{|I|=p} \sum_{|J|=q} f_{I,J}\; dz_I \wedge d\bar{z}_J ,$$

where the coefficients $f_{I,J}$ are functions or distributions in $\mathscr{D}$. We use standard multi–index notation: thus the *multi–index* I is an ordered set of integers $1 \leq i_1 < ... < i_p \leq n$; $|I|$ is its *length*, here p ; $dz_I = dz_{i_1} \wedge ... \wedge dz_{i_p}$. We shall systematically use the notation $dz = dz_1 \wedge ... \wedge dz_n$.

We denote by $\mathcal{D}\,'(\mathscr{D}; \Lambda^{p,q})$ the space of all currents (I.1.1); $f \in C^\infty(\mathscr{D}; \Lambda^{p,q})$ if the coefficients $f_{I,J}$ are C^∞ functions in $\mathscr{D}$; $f \in C^\infty(\mathscr{C}\!\ell\,\mathscr{D}; \Lambda^{p,q})$ if they extend as C^∞ functions to the closure of $\mathscr{D}$. These spaces carry natural topologies (see [Schw]). We have:

$$(\text{I.1.2}) \qquad \bar{\partial}_z f = \sum_{|I|=p} \sum_{|J|=q} \sum_{k=1}^{n} (\partial f_{I,J}/\partial \bar{z}_k)\; d\bar{z}_k \wedge dz_I \wedge d\bar{z}_J .$$

These formulas define, for each p , $0 \leq p \leq n$, the differential complex

$$(\text{I.1.3}) \qquad \bar{\partial}_z : \mathcal{D}\,'(\mathscr{D}; \Lambda^{p,q}) \to \mathcal{D}\,'(\mathscr{D}; \Lambda^{p,q+1}) ,\quad q = 0,1,...,$$

and its subcomplex,

$$(\text{I.1.4}) \qquad \bar{\partial}_z : C^\infty(\mathscr{D}; \Lambda^{p,q}) \to C^\infty(\mathscr{D}; \Lambda^{p,q+1}) ,\quad q = 0,1,....$$

We shall refer to (I.1.3) or (I.1.4) as the *Cauchy–Riemann* (or *Dolbeault*) *complex*, of degree p , on $\mathbb{C}^n$.

It is convenient to deal with the case $p = n$. Then (I.1.1) reads

$$(I.1.5) \qquad\qquad f = dz \wedge f_0 \quad , \quad f_0 = \sum_{|J|=q} f_J \, d\bar{z}_J \, ,$$

and we have

$$(I.1.6) \qquad\qquad \bar{\partial}_z f = df \, ,$$

so that (I.1.3) & (I.1.4) can be viewed as subcomplexes of the De Rham complex.

We denote by (z, z') the variable point in $\mathbb{C}^n \times \mathbb{C}^n$. In the sequel, and until specified otherwise, d will stand for the exterior derivative in z–space, d' for the exterior derivative in z'–space. We shall also use the notation

$$(I.1.7) \qquad\qquad D = d + d' \, ;$$

D is the exterior derivative in (z, z')–space.

REMARK I.1.1.– In accordance with the notation just introduced, the differential of a coordinate function z'_j ought to be denoted by $d'z'_j$. We shall, however, write dz'_j and $d\bar{z}'_j$. Generally speaking we shall use the notation d' whenever this serves the purpose of making it clear that the exterior derivative is taken with respect to the variables z'. ❾

One of our basic data will be a C^∞ map

$$(I.1.8) \qquad\qquad a = (a_1, \ldots, a_n) : \mathscr{D} \times \mathscr{D} \to \mathbb{C}^n \, .$$

We shall make use of the following differential operators (of order 1)

(I.1.9)

$$d^{a} = d + d[a(z,z')\cdot(z-z')]\wedge \quad , \quad d'^{a} = d' + d'[a(z,z')\cdot(z-z')]\wedge .$$

Obviously,

(I.1.10)

$$d^{a} = e^{-a\cdot(z-z')} d e^{a\cdot(z-z')} .$$

We shall also use the notation

(I.1.11)

$$D^{a} = d^{a} + d'^{a} .$$

The differential operators d^{a}, d'^{a} and D^{a} will act on differential forms in $\mathscr{D} \times \mathscr{D}$.

Since we shall make frequent use of the difference vector $z-z'$ it is convenient to denote it by a single letter; we shall use $\lambda = z-z'$. In the sequel we are going to avail ourselves of the following obvious assertion:

LEMMA I.1.1.– *If a C^{∞} form F in $\mathbb{C}^{n}\times\mathbb{C}^{n}$ is the pull–back of a C^{∞} form f in $\mathbb{C}^{n}$ under the map $(z,z') \to z-z'$ then DF is the pull–back of df.*

We must explain how we use *partial integration of differential forms* in (z,z')–space. For us partial integration of such a form will always mean its integration on a chain in z'–space. Let F denote a differential form in $\mathscr{D} \times \mathscr{D}$ of bidegree (p,q) with respect to z :

(I.1.12)

$$F = \sum_{|I|=p} \sum_{|J|=q} \sum_{\nu=0}^{2n} F_{I,J}^{(\nu)} \wedge dz_{I} \wedge d\bar{z}_{J} ,$$

where the $F_{I,J}^{(\nu)}$ are differential forms of total degree ν with respect to z' whose coefficients are, say, continuous functions of (z, z') in $\mathcal{D} \times \mathcal{D}$. Let then c be a compact r–chain in $\mathcal{D}$. We shall then have, *by definition*,

$$(\text{I.1.13}) \qquad \int_c F = \sum_{|I|=p} \sum_{|J|=q} \left\{ \int_c F_{I,J}^{(r)} \right\} dz_I \wedge d\bar{z}_J .$$

The integral $\int_c F_{I,J}^{(r)}$ is that of an r–form in z'–space which depend on $z \in \mathcal{D}$.

An important consequence of Definition (I.1.13) is Stokes' theorem and the formula of differentiation under the integral sign,

$$(\text{I.1.14}) \qquad \int_c d'F = \int_{\partial c} F \quad , \quad d \int_c F = (-1)^r \int_c dF .$$

In applying (I.1.14) keep in mind that d (resp., d') stands for the exterior derivative in z–space (resp., z'–space).

In the actual computations of the integrals over open subsets of $\mathcal{D}$ we shall always choose the orientation of $\mathbb{C}^n$ in which the *positive volume form* is equal to

$$(\text{I.1.15}) \qquad (2\imath)^{-n} d\bar{z} \wedge dz = dx_1 \wedge \ldots \wedge dx_n \wedge dy_1 \wedge \ldots \wedge dy_n .$$

Concerning notation we must go back to multi–indices.

Let $\omega = (\omega_1, \ldots, \omega_n)$ be a set of n one–forms in $\mathcal{D}$ and let $J = (j_1, \ldots, j_p)$ be any multi–index of length p, $1 \leq j_1 < \ldots < j_p \leq n$. It is agreed that

$$\omega_J = \omega_{j_1} \wedge \ldots \wedge \omega_{j_p} .$$

By J^* we shall mean the complement $[1, \ldots, n] \backslash J$; J^* will be ordered. *i. e.*, strictly increasing. It is then convenient to use the following notation:

$$(\text{I.1.16}) \qquad \omega_{(J)} = (-1)^{j_1 + \ldots + j_p - p} \omega_{J^*} .$$

In particular,

$$\omega_{(J)} = (-1)^{n(n-1)/2} \quad if \quad J = [1,...,n].$$

Note also that we have

(I.1.17) $$\omega_J \wedge \omega_{(J)} = (-1)^{p(p-1)/2} \omega_1 \wedge ... \wedge \omega_n .$$

If K is another ordered set of q integers $k \in [1,...,n]$, we write $JK = \{j_1,...,j_p, k_1,...,k_q\}$; in general JK is not ordered and might contain repetitions. If J is not ordered but does not contain repetitions, $[J]$ will stand for the multi–index obtained by ordering J ; $\epsilon(J)$ shall denote the signum of the permutation that transforms J into $[J]$: $\epsilon(J) = 1$ if the permutation is even, $\epsilon(J) = -1$ if it is odd. If j is the νth element of the set of integers J , we write $J\backslash j$ for the complement of j in J and thus $\epsilon(j(J\backslash j)) = (-1)^{\nu-1}$.

Thus, if j is an element of J we have

(I.1.18) $$\omega_j \wedge \omega_{(J)} = \epsilon(j(J\backslash j)) \omega_{(J\backslash j)} .$$

I.2. BOCHNER–MARTINELLI FORMULA WITH EXPONENTIAL

In the first part of the present section we shall reason in $\mathbb{R}^{2n} \cong \mathbb{C}^n$ where the complex variable is called λ . Until specified otherwise, by $\bar{\partial}_j$ we shall mean $\partial/\partial\bar{\lambda}_j$; the exterior derivative shall be denoted by d . The following distributions in $\mathbb{R}^{2n}$ will play a central role in the forthcoming:

$$(\text{I.2.1}) \qquad \mathcal{E}_j^{(0)}(\lambda) = (n-1)! \; \bar{\lambda}_j/|\lambda|^{2n} \; ,$$

$$(\text{I.2.2}) \qquad \mathcal{E}_j^{(r)}(\lambda) = (-1)^r(n-r-1)! \; \bar{\lambda}_j/|\lambda|^{2(n-r)} \quad if \; 1 \leq r \leq n-1 \; ;$$

the notation $\mathcal{E}_j^{(r)}$ will mean $\mathcal{E}_j^{(r)}(\lambda)$.

PROPOSITION I.2.1.– *Let δ denote the Dirac distribution in $\mathbb{R}^{2n}$. We have*

$$(\text{I.2.3}) \qquad \sum_{j=1}^{n} \bar{\partial}_j \mathcal{E}_j^{(0)} = \pi^n \delta \; .$$

Proof : When $n = 1$ formula (I.2.3) is a restatement of the fact that $1/\pi\lambda$ is a fundamental solution of $\partial/\partial\bar{\lambda}$ in $\mathbb{R}^2$. When $n > 1$, it follows from the fact that $-(n-2)!/(4\pi^n|\lambda|^{2(n-1)})$ is a fundamental solution of the Laplacian in λ–space $\mathbb{R}^{2n}$. ∎

When $1 \leq r \leq n-1$ we have:

$$(\text{I.2.4}) \qquad \bar{\partial}_k \mathcal{E}_j^{(r)} = \lambda_k \mathcal{E}_j^{(r-1)} + \delta_{jk}\Delta^{(r)},$$

where δ_{jk} is the Kronecker index and

$$(\text{I.2.5}) \qquad \Delta^{(r)} = (-1)^r(n-r-1)!/|\lambda|^{2(n-r)} \; .$$

Also notice that

$$(I.2.6) \qquad \sum_{j=1}^{n} \lambda_j \mathscr{E}_j^{(r)} = -(n-r-1)\Delta^{(r+1)} \qquad (0 \le r \le n-2);$$

$$(I.2.7) \qquad \sum_{j=1}^{n} \lambda_j \mathscr{E}_j^{(n-1)} = (-1)^{n-1}.$$

At this point we switch to (z,z')–space, precisely to $\mathscr{D} \times \mathscr{D}$. Henceforth λ takes on the meaning $z - z'$. We shall be using the forms $D\lambda_j = dz_j - dz'_j$, $D a_j$ ($j = 1,...,n$) and the concomitant notation introduced in Sect. 1 : e. g., $D a_J = D a_{j_1} \wedge$ ...$\wedge D a_{j_p}$ if $J = (j_1,...,j_p)$. We define the following $(2n-1)$–forms in $\mathscr{D} \times \mathscr{D}$:

$$(I.2.8) \; E_a^{(\nu)} = \sum_{|J|=\nu+1} \sum_{j \in J} \epsilon(j(J \setminus j)) \mathscr{E}_j^{(\nu)} (D a)_{J \setminus j} \wedge D\bar{\lambda}_{(J)} \wedge D\lambda, \qquad \nu = 0,1,..,n-1,$$

$$(I.2.9) \qquad E_a = \sum_{\nu=0}^{n-1} (-1)^{\nu(\nu+1)/2} E_a^{(\nu)}.$$

Note in passing that

$$(I.2.10) \qquad E_a^{(0)} = E^{(0)} = \sum_{j=1}^{n} \mathscr{E}_j^{(0)} D\bar{\lambda}_{(j)} \wedge D\lambda.$$

We recall that

$$(I.2.11) \qquad D\lambda_{(j)} = (-1)^{j-1} D\lambda_1 \wedge...\wedge D\lambda_{j-1} \wedge D\lambda_{j+1} \wedge...\wedge D\lambda_n.$$

Formula $(I.2.3)$ can be rewritten as follows:

(I.2.12) $$DE^{(0)} = \pi^n \delta(\lambda) D\bar{\lambda} \wedge D\lambda$$

where now $\delta(\lambda)$ must be regarded as the Dirac measure on the diagonal $z = z'$. Below we write $D\,\mathbf{a} = D\,\mathbf{a}_1 \wedge ... \wedge D\,\mathbf{a}_n$. We state

THEOREM I.2.1.– *We have*

(I.2.13) $$D^{\mathbf{a}}E_{\mathbf{a}} = \left\{ \pi^n \delta(\lambda) D\bar{\lambda} - (-1)^n D\,\mathbf{a} \right\} \wedge D\lambda .$$

Proof : We seek a convenient expression of $DE_{\mathbf{a}}^{(\nu)}$ when $1 \leq \nu \leq n-1$. From Lemma I.1.1 and from (I.2.8) we derive

$$DE_{\mathbf{a}}^{(\nu)} = \sum_{k=1}^{n} \sum_{|J|=\nu+1} \sum_{j \in J} \epsilon(j(J\backslash j)) \bar{\partial}_k \mathscr{E}_j^{(\nu)} \, D\bar{\lambda}_k \wedge (D\,\mathbf{a})_{J\backslash j} \wedge D\bar{\lambda}_{(J)} \wedge D\lambda$$

$$= (-1)^\nu \sum_{|J|=\nu+1} \sum_{j,k \in J} \epsilon(j(J\backslash j)) \bar{\partial}_k \mathscr{E}_j^{(\nu)} \, (D\,\mathbf{a})_{J\backslash j} \wedge D\bar{\lambda}_k \wedge D\bar{\lambda}_{(J)} \wedge D\lambda .$$

We apply (I.2.3) and (I.2.4):

$$(-1)^\nu DE_{\mathbf{a}}^{(\nu)} =$$

$$\sum_{|J|=\nu+1} \sum_{j,k \in J} \epsilon(j(J\backslash j)) \epsilon(k(J\backslash k)) \bar{\partial}_k \mathscr{E}_j^{(\nu)} \, (D\,\mathbf{a})_{J\backslash j} \wedge D\bar{\lambda}_{(J\backslash k)} \wedge D\lambda =$$

$$\sum_{|J|=\nu+1} \sum_{j \in J} \left\{ [\Delta^{(\nu)} + \lambda_j \mathscr{E}_j^{(\nu-1)}](D\,\mathbf{a})_{J\backslash j} \wedge D\bar{\lambda}_{(J\backslash j)} + \right.$$

$$\left. \sum_{k \in J\backslash j} \epsilon(j(J\backslash j)) \epsilon(k(J\backslash k)) \lambda_k \mathscr{E}_j^{(\nu-1)} \, (D\,\mathbf{a})_{J\backslash j} \wedge D\bar{\lambda}_{(J\backslash k)} \right\} \wedge D\lambda .$$

But, of course,

$$\sum_{j\in J}\sum_{k\in J\setminus j}\epsilon(j(J\setminus j))\epsilon(k(J\setminus k))\lambda_k\mathscr{E}_j^{(\nu-1)}\,(D\alpha)_{J\setminus j}\wedge D\overline{\lambda}_{(J\setminus k)} =$$

$$\sum_{j\in J}\sum_{k\in J\setminus j}\epsilon(j(J\setminus j))\epsilon(k(J\setminus k))\lambda_j\mathscr{E}_k^{(\nu-1)}\,(D\alpha)_{J\setminus k}\wedge D\overline{\lambda}_{(J\setminus j)}$$

whence

$$(-1)^{\nu}DE_{\alpha}^{(\nu)} = \sum_{|J|=\nu+1}\sum_{j\in J}\left\{[\Delta^{(\nu)}+\lambda_j\mathscr{E}_j^{(\nu-1)}](D\alpha)_{J\setminus j}\wedge D\overline{\lambda}_{(J\setminus j)} + \right.$$

$$\left. \sum_{k\in J\setminus j}\epsilon(j(J\setminus j))\epsilon(k(J\setminus k))\lambda_j\mathscr{E}_k^{(\nu-1)}\,(D\alpha)_{J\setminus k}\right\}\wedge D\overline{\lambda}_{(J\setminus j)}\wedge D\lambda \ .$$

We observe that the map $(J,j) \to (J\setminus j,j)$ is a bijection of the set of pairs $\{\ (J,j)\ ;\ |J| = r + 1\ ,\ j \in J\ \}$ onto the set of pairs $\{\ (I,\ell)\ ;\ |I| = r\ ,\ \ell \notin I\ \}$. We obtain

(I.2.14)

$$(-1)^{\nu}DE_{\alpha}^{(\nu)} = \sum_{|I|=\nu}\sum_{\ell\notin I}\left\{[\Delta^{(\nu)} + \lambda_\ell\mathscr{E}_\ell^{(\nu-1)}](D\alpha)_I + \right.$$

$$\left. \sum_{i\in I}\epsilon(\ell I)\epsilon(i([\ell I]\setminus i))\epsilon(\ell(I\setminus i))\mathscr{E}_i^{(\nu-1)}(D\alpha_\ell\lambda_\ell)\wedge(D\alpha)_{I\setminus i}\right\}\wedge D\overline{\lambda}_{(I)}\wedge D\lambda \ .$$

We note that

$$\sum_{\ell\notin I}[\Delta^{(\nu)}+\lambda_\ell\mathscr{E}_\ell^{(\nu-1)}] = (n-\nu)\Delta^{(\nu)} + \sum_{\ell=1}^{n}\lambda_\ell\mathscr{E}_\ell^{(\nu-1)} - \sum_{i\in I}\lambda_i\mathscr{E}_i^{(\nu-1)} =$$

$$- \sum_{i\in I}\lambda_i\mathscr{E}_i^{(\nu-1)}$$

by (I.2.6) (notice that $\nu - 1 \leq n - 2$). Thus

(I.2.15)

$$\sum_{\ell \notin I} [\Delta^{(\nu)} + \lambda_\ell \mathcal{E}_\ell^{(\nu-1)}](D\alpha)_I = - \sum_{i \in I} \epsilon(i(\Lambda i) \mathcal{E}_i^{(\nu-1)}(\lambda_i D\alpha_i) \wedge (D\alpha)_{\Lambda i} .$$

On the other hand, it is readily checked that

(I.2.16) $$\epsilon(\ell I)\epsilon(i([\ell I]\backslash i))\epsilon(\ell(\Lambda i)) = - \epsilon(i(\Lambda i)) .$$

By combining (I.2.14), (I.2.15) and (I.2.16) we get

$$(-1)^\nu DE_\alpha^{(\nu)} = - \sum_{|I|=\nu} \sum_{i \in I} \epsilon(i(\Lambda i) \mathcal{E}_i^{(\nu-1)} (D\alpha \cdot \lambda) \wedge (D\alpha)_{\Lambda i} \wedge D\bar{\lambda}_{(I)} \wedge D\lambda ,$$

that is to say,

$$(-1)^\nu DE_\alpha^{(\nu)} = - (D\alpha \cdot \lambda) \wedge E_\alpha^{(\nu-1)} .$$

or equivalently, after multiplying both sides by $(-1)^{\nu(\nu-1)/2}$,

(I.2.17) $$(-1)^{\nu(\nu+1)/2} DE_\alpha^{(\nu)} = - (-1)^{\nu(\nu-1)/2}(D\alpha \cdot \lambda) \wedge E_\alpha^{(\nu-1)} ,$$

which is valid for $1 \leq \nu \leq n - 1$.

We combine (I.2.12) and (I.2.17):

(I.2.18) $$DE_\alpha + (D\alpha \cdot \lambda) \wedge E_\alpha =$$

$$\pi^n \delta(\lambda) D\bar{\lambda} \wedge D\lambda + (-1)^{n(n-1)/2}(D\alpha \cdot \lambda) \wedge E_\alpha^{(n-1)} .$$

(I.2.19)
$$D\alpha \cdot \lambda = \sum_{\ell=1}^{n} \lambda_\ell D\alpha_\ell .$$

Now, according to (I.2.8),

$$E_\alpha^{(n-1)} = (-1)^{n(n-1)/2} \sum_{j=1}^{n} \mathscr{E}_j^{(n-1)} (D\alpha)_{(j)} \wedge D\lambda .$$

It follows that

$$(-1)^{n(n-1)/2}(D\alpha \cdot \lambda) \wedge E_\alpha^{(n-1)} = \sum_{j=1}^{n} \lambda_j \mathscr{E}_j^{(n-1)} D\alpha \wedge D\lambda = (-1)^{n-1} D\alpha \wedge D\lambda$$

by (I.2.7). Combining this with (I.2.18) yields Th. I.2.1. ∎

Let now Ω denote a relatively compact open subset of the domain $\mathscr{D} \subset \mathbf{C}^n$ with C^∞ boundary $\partial\Omega$. Let q be an integer, $0 \leq q \leq n$, f an arbitrary element of $C^\infty(\mathscr{C}\ell\,\Omega; \Lambda^{n,q})$: f has the expression (I.1.5) with coefficients $f_J \in C^\infty(\mathscr{C}\ell\,\Omega)$. We write $f(z)$ (resp., $f(z')$) to indicate that f is a form in z–space (resp., z'–space). Below we use the notation

(I.2.20) $$K_\Omega^\alpha f(z) = (-2\iota\pi)^{-n} \int_\Omega e^{\alpha(z,z')\cdot(z-z')} f(z') \wedge E_\alpha(z,z') ;$$

(I.2.21) $$\mathcal{R}_\Omega^\alpha f(z) = (-2\iota\pi)^{-n} \int_\Omega e^{\alpha(z,z')\cdot(z-z')} f(z') \wedge D\alpha(z,z') \wedge D\lambda .$$

In (I.2.20) and (I.2.21) we are using partial integration with respect to z' (see Sect. 1). We shall also use the notation $K_{\partial\Omega}^\alpha f$: its definition differs from (I.2.20) only in that the integration with respect to z' is carried out over the boundary $\partial\Omega$ of Ω . In (I.2.21) we could have substituted dz for $D\lambda$ since the form $f(z')$ contains a factor $\mathrm{d}z'$. Note also that $\mathcal{K}_\Omega^a f \equiv 0$ when $q = 0$ (by the definition of E_a).

COROLLARY I.2.1.– *Suppose* $0 \leq q \leq n$. *For any* $f \in C^\infty(\mathscr{C}\mathscr{L}\Omega;\Lambda^{n,q})$ *we have, in the open set* Ω ,

$$(\mathrm{I.2.22}) \qquad (2\imath\pi)^{-n} \int_\Omega e^{\,a\,(z,z')\cdot(z-z')} f\,(z') \wedge \mathrm{D}^{\,a} E_a\,(z,z') = f\,(z) - \mathcal{R}_\Omega^a f\,(z) \ .$$

Proof: By (I.2.13) and (I.2.21) it suffices to show that

$$(\mathrm{I.2.23}) \qquad\qquad f\,(z) = (2\imath)^{-n} \int_\Omega f\,(z') \wedge \delta(z-z') \mathrm{d}(\bar{z}-\bar{z}') \wedge \mathrm{d}(z-z') \ .$$

In order to prove (I.2.23) it suffices to consider the case where

$$f\,(z) = f_J(z) \mathrm{d}\bar{z}_J \wedge \mathrm{d}z \ , \quad |J| = q \ .$$

Then

$$\int_\Omega f\,(z') \wedge \delta(z-z') \mathrm{d}(\bar{z}-\bar{z}') \wedge \mathrm{d}(z-z') = f_J(z) \int_{\mathbb{R}^{2n}} \delta(z-z') \mathrm{d}\bar{z}'_J \wedge \mathrm{d}z' \wedge \mathrm{d}(\bar{z}-\bar{z}') \wedge \mathrm{d}z \ .$$

We apply (I.1.17) twice; for the purpose of integration in z'–space, we have

$$\mathrm{d}\bar{z}'_J \wedge \mathrm{d}z' \wedge \mathrm{d}(\bar{z}-\bar{z}') = (-1)^{q(q-1)/2} \mathrm{d}\bar{z}'_J \wedge \mathrm{d}z' \wedge \mathrm{d}(\bar{z}-\bar{z}')_J \wedge \mathrm{d}(\bar{z}-\bar{z}')_{(J)} =$$

$$(-1)^{q(q-1)/2+n-q} \mathrm{d}\bar{z}'_J \wedge \mathrm{d}z' \wedge \mathrm{d}\bar{z}_J \wedge \mathrm{d}\bar{z}'_{(J)} = (-1)^{q(q-1)/2} \mathrm{d}\bar{z}'_J \wedge \mathrm{d}\bar{z}'_{(J)} \wedge \mathrm{d}z' \wedge \mathrm{d}\bar{z}_J =$$

$$\mathrm{d}\bar{z}' \wedge \mathrm{d}z' \wedge \mathrm{d}\bar{z}_J$$

whence (I.2.23). ☻

THEOREM I.2.2.– *Suppose* $0 \leq q \leq n$. *For any* $f \in C^{\infty}(\mathscr{C} \mathscr{L} \Omega; \Lambda^{n,q})$ *we have, in the open set* Ω ,

(I.2.24) $d[(-1)^q \mathcal{K}_{\Omega}^{a} f] + (-1)^{q+1} \mathcal{K}_{\Omega}^{a} df = f - \mathcal{I}_{\Omega}^{a} f - (-1)^q \mathcal{K}_{\partial\Omega}^{a} f$.

Proof: By virtue of (I.1.10), (I.1.11) and of the expression (I.2.20) we may write:

$$(-2\imath\pi)^n d\mathcal{K}_{\Omega}^{a} f(z) =$$

$$(-1)^{n+q} \int_{\Omega} e^{a(z,z') \cdot (z-z')} f(z') \wedge \left\{ D^{a} E_{a}(z,z') - d'^{a} E_{a}(z,z') \right\} .$$

But by the Stokes theorem,

$$(-1)^{n+q} \int_{\Omega} e^{a(z,z') \cdot (z-z')} f(z') \wedge d'^{a} E_{a}(z,z') =$$

$$- \int_{\Omega} e^{a(z,z') \cdot (z-z')} df(z') \wedge E_{a}(z,z') + \int_{\partial\Omega} e^{a(z,z') \cdot (z-z')} f(z') \wedge E_{a}(z,z') ,$$

whence

$$(2\imath\pi)^n (d\mathcal{K}_{\Omega}^{a} f - \mathcal{K}_{\Omega}^{a} df)(z) =$$

$$(-1)^q \int_{\Omega} e^{a(z,z') \cdot (z-z')} f(z') \wedge D^{a} E_{a}(z,z') - (2\imath\pi)^n \mathcal{K}_{\partial\Omega}^{a} f(z) .$$

Cor. I.2.1 implies (I.2.24). ✊

COROLLARY I.2.2.– *For any* $f \in C^{\infty}(\mathscr{C} \mathscr{L} \Omega; \Lambda^{n,0})$ *we have, in the open set* Ω ,

(I.2.25) $f = - \mathcal{K}_{\Omega}^{a} df + \mathcal{I}_{\Omega}^{a} f + \mathcal{K}_{\partial\Omega}^{a} f$.

$\mathcal{Proof}$: Indeed, $K_\Omega^a f \equiv 0$ when $q = 0$. ☻

COROLLARY I.2.3.– *For any* $f \in C^\infty(\mathscr{C}\!\ell\,\Omega; \Lambda^{n,n})$ *we have, in the open set* Ω ,

$$(I.2.26) \qquad\qquad f = (-1)^n \mathrm{d}K_\Omega^a f + \imath_\Omega^a f \ .$$

$\mathcal{Proof}$: Indeed, if the degree of the form f is equal to $2n$ then $\mathrm{d}f \equiv 0$ and the integral of f over the $(2n-1)$–chain $\partial\Omega$ is perforce equal to zero. ☻

I.3. KOPPELMAN FORMULAS WITH EXPONENTIAL

In the present section we shall reason under the hypothesis that *the open set* $\Omega \subset\subset \mathscr{D} \subset \mathbb{C}^n$ *is convex* and has a boundary $\partial\Omega$ of class $\mathcal{C}^\infty$, that is, we shall assume that there is a real–valued $\mathcal{C}^\infty$ function ρ in an open neighborhood $\tilde{\Omega}$ of $\mathscr{Cl}$ Ω in $\mathbb{C}^n$ such that $\Omega = \{ z \in \tilde{\Omega} \; ; \; \rho(z) < 0 \}$, $d\rho \neq 0$ at every point of $\partial\Omega$ and

$$(I.3.1) \qquad \mathscr{R}e\,[\partial\rho(z')\cdot\lambda] < 0 \; , \;\; \forall \; z \in \Omega, \, z' \in \partial\Omega \; .$$

Here and in the sequel $\partial\rho = (\partial_1\rho,...,\partial_n\rho)$, $\partial_j\rho = \partial\rho/\partial z_j$. We recall that $\lambda = z - z'$.

We introduce an additional variable, t , varying in $[0,1]$. Here we shall make use of the letter $\mathbb{D}$ to denote the exterior derivative in (z,z',t)–space: $\mathbb{D} = D + d_t = d + d' + d_t$. However, we write dz, dz', $D\lambda$ $(= dz - dz')$, dt , rather than $\mathbb{D}z$, $\mathbb{D}z'$, $\mathbb{D}\lambda$, $\mathbb{D}t$. The map a will depend also on t and we shall write $\mathbb{D}^a = \mathbb{D} + \mathbb{D}(a \cdot \lambda)\wedge$.

Consider the following function, defined and $\mathcal{C}^\infty$ in $\Omega\times\partial\Omega\times[0,1]$ and valued in $\mathbb{C}^n$:

$$\zeta = t\partial\rho(z')/[\partial\rho(z')\cdot\lambda] + (1-t)\bar{\lambda}/|\lambda|^2 \; .$$

In $\Omega\times\partial\Omega\times[0,1]$ we have

$$(I.3.2) \qquad\qquad\qquad \lambda\cdot\zeta \equiv 1$$

and, as a consequence,

$$(I.3.3) \qquad\qquad\qquad \sum_{j=1}^{n}\lambda_j\mathbb{D}\bar{\zeta}_j\wedge D\lambda = 0 \; .$$

$\mathscr{Proof}$ $\mathscr{of}$ (I.3.3): From (I.3.2) we derive

$$\sum_{j=1}^{n} \lambda_j \mathbb{D}\overline{\zeta}_j + \overline{\zeta}_j D\lambda_j = 0 \; .$$

It suffices to multiply the left–hand side of this identity by $D\lambda$ to derive (I.3.3). ☻

An immediate consequence of (I.3.3) is that we also have, in $\Omega \times \partial\Omega \times [0,1]$,

(I.3.4) $$\mathbb{D}\overline{\zeta} \wedge D\lambda = 0 \; .$$

We define the following differential form of degree $2n-1$ in $\Omega \times \partial\Omega \times [0,1]$:

(I.3.5) $$F_a^{(0)} = F^{(0)} = (n-1)! \sum_{j=1}^{n} \overline{\zeta}_j \mathbb{D}\overline{\zeta}_{(j)} \wedge D\lambda \; .$$

In the present section the map a shall depend on $t \in [0,1]$ in C^{∞} fashion. We write then, for $1 \leq \nu \leq n - 1$,

(I.3.6) $$F_a^{(\nu)} = (-1)^{\nu}(n-\nu-1)! \sum_{|J|=\nu+1} \sum_{j \in J} \epsilon(\mathcal{j}(J \backslash \mathcal{j}))\overline{\zeta}_j (\mathbb{D} a)_{J \backslash \mathcal{j}} \wedge (\mathbb{D}\overline{\zeta})_{(J)} \wedge D\lambda \; .$$

Lastly we define

(I.3.7) $$F_a = \sum_{\nu=0}^{n-1} (-1)^{\nu(\nu+1)/2} \; F_a^{(\nu)} \; .$$

THEOREM I.3.1.– *In $\Omega \times \partial\Omega \times [0,1]$ we have*

(I.3.8) $$\mathbb{D}^a F_a = (-1)^{n-1} \mathbb{D} a \wedge D\lambda \; .$$

Proof: We have $\mathbb{D}F_a^{(0)} = 0$ in $\Omega\times\partial\Omega\times[0,1]$ by (I.3.4) and (I.3.5). Suppose $1 \leq \nu \leq n-1$. We have

$$\mathbb{D}F_a^{(\nu)} = (n-\nu-1)! \sum_{|J|=\nu+1} \sum_{j\in J} (\mathbb{D}a)_{J\backslash j}\wedge(\mathbb{D}\overline{\zeta})_{(J\backslash j)}\wedge D\lambda ,$$

which is to say

(I.3.9) $$\mathbb{D}F_a^{(\nu)} = (n-\nu)! \sum_{|J|=\nu} (\mathbb{D}a)_J\wedge(\mathbb{D}\overline{\zeta})_{(J)}\wedge D\lambda .$$

On the other hand, for any ν, $0 \leq \nu \leq n-1$,

$$(-1)^\nu(\mathbb{D}a\,\lambda)\wedge F_a^{(\nu)}/(n-\nu-1)! = \sum_{|J|=\nu+1} \sum_{j\in J}\left\{\lambda_j\overline{\zeta}_j(\mathbb{D}a)_J\wedge(\mathbb{D}\overline{\zeta})_{(J)} + \right.$$

$$\left.\epsilon(j(J\backslash j))\sum_{k\notin J}\lambda_k\overline{\zeta}_j\mathbb{D}a_k\wedge(\mathbb{D}a)_{J\backslash j}\wedge(\mathbb{D}\overline{\zeta})_{(J)}\right\}\wedge D\lambda .$$

By virtue of (I.3.2) we get

(I.3.10) $$(-1)^\nu(\mathbb{D}a\cdot\lambda)\wedge F_a^{(\nu)}/(n-\nu-1)! =$$

$$\sum_{|J|=\nu+1}\left\{(\mathbb{D}a)_J - \sum_{k\notin J}\lambda_k\left\{\overline{\zeta}_k(\mathbb{D}a)_J - \sum_{j\in J}\epsilon(j(J\backslash j))\overline{\zeta}_j\mathbb{D}a_k\wedge(\mathbb{D}a)_{J\backslash j}\right\}\right\}\wedge(\mathbb{D}\overline{\zeta})_{(J)}\wedge D\lambda .$$

If $K = [k(J\backslash j)]$ we have $|K| = \nu + 1$ and $K\backslash k = J\backslash j$. Then

$$\mathbb{D}a_k\wedge(\mathbb{D}a)_{J\backslash j} = \epsilon(k(K\backslash k))(\mathbb{D}a)_K , \quad \epsilon(j(J\backslash j))\epsilon(k(K\backslash k)) = -\epsilon(jK)\epsilon(kJ) ,$$

and therefore

$$\sum_{|J|=\nu+1} \sum_{k\notin J} \lambda_k \Big\{ \bar\zeta_k (\mathbb{D}a)_J - \sum_{j\in J} \epsilon(j(J\backslash j))\bar\zeta_j \mathbb{D}\, a_k \wedge (\mathbb{D}a)_{J\backslash j} \Big\} \wedge (\mathbb{D}\bar\zeta)_{(J)} =$$

$$\sum_{|J|=\nu+1} \sum_{k\notin J} \lambda_k \Big\{ \bar\zeta_k (\mathbb{D}a)_J + \sum_{j\in J} \epsilon(jK)\epsilon(kJ))\bar\zeta_j \mathbb{D}\, a_K \Big\} \wedge (\mathbb{D}\bar\zeta)_{(J)}$$

where, as before, $K = [kJ]\backslash j$. Let us avail ourselves of (I.3.3): by carrying out the exterior product of the left–hand side in (I.3.3) with $(\mathbb{D}\bar\zeta)_{([kJ])}$ we obtain:

$$\epsilon(kJ)\lambda_k (\mathbb{D}\bar\zeta)_{(J)} \wedge \mathrm{D}\lambda = -\sum_{j\in J} \epsilon(j([kJ]\backslash j))\lambda_j (\mathbb{D}\bar\zeta)_{([kJ]\backslash j)} \wedge \mathrm{D}\lambda \ .$$

Thus

$$\sum_{|J|=\nu+1} \sum_{k\notin J} \lambda_k \Big\{ \bar\zeta_k (\mathbb{D}a)_J - \sum_{j\in J} \epsilon(j(J\backslash j))\bar\zeta_j \mathbb{D}\, a_k \wedge (\mathbb{D}a)_{J\backslash j} \Big\} \wedge (\mathbb{D}\bar\zeta)_{(J)} \wedge \mathrm{D}\lambda =$$

$$\sum_{|J|=\nu+1} \sum_{j\in J} \sum_{k\notin J} \epsilon(jK)\epsilon(kJ))\lambda_k \bar\zeta_j \mathbb{D}\, a_K \wedge (\mathbb{D}\bar\zeta)_{(J)} \wedge \mathrm{D}\lambda -$$

$$\sum_{|J|=\nu+1} \sum_{k\notin J} \sum_{j\in J} \epsilon(jK)\epsilon(kJ))\lambda_j \bar\zeta_k \mathbb{D}\, a_J \wedge (\mathbb{D}\bar\zeta)_{(K)} \wedge \mathrm{D}\lambda \equiv 0$$

since the map $(J,j,k) \to (K,k,j)$, with $K = [k(J\backslash j)]$, is a permutation of the set of triplets $\{ \ (J,j,k) \ ; \ |J| = \nu+1 \ , \ j \in J, \ k \notin K\}$. Taking this into account in (I.3.10) yields

$$(-1)^\nu (\mathbb{D}a \cdot \lambda) \wedge F_a^{(\nu)}/(n-\nu-1)! = \sum_{|J|=\nu+1} (\mathbb{D}a)_J \wedge (\mathbb{D}\bar\zeta)_{(J)} \wedge \mathrm{D}\lambda \ .$$

Combining this with (I.3.9) yields

$$(-1)^{\nu(\nu+1)/2} \mathbb{D}\, a_{F}^{(\nu)} = (-1)^{\nu(\nu+1)/2} (n-\nu)! \sum_{|J|=\nu} (\mathbb{D}a)_J \wedge (\mathbb{D}\bar\zeta)_{(J)} \wedge \mathrm{D}\lambda -$$

$$(-1)^{(\nu+1)(\nu+2)/2}(n-\nu-1)! \sum_{|J|=\nu+1} (\mathbb{D}\,a\,)_J \wedge (\mathbb{D}\overline{\zeta})_{(J)} \wedge D\lambda$$

since $\nu+\nu(\nu+1)/2 = (\nu+1)(\nu+2)/2 + 1 \bmod 2$. Th. I.3.1 follows at once from this and from the definition of F_a , (I.3.7). ☻

The proof of the next statement follows directly by computation:

LEMMA I.3.1.– *Assume* $1 \leq r \leq n$. *Let* ψ , u_1 ,..., u_r *be* $r+1$ C^∞ *functions. We have*

(I.3.11)
$$\sum_{j=1}^{r}(-1)^{j-1}\psi u_j \mathrm{d}(\psi u_1)\wedge...\wedge \mathrm{d}(\widehat{\psi u_j})\wedge...\wedge \mathrm{d}(\psi u_r) =$$

$$\psi^r \sum_{j=1}^{r}(-1)^{j-1} u_j \mathrm{d}u_1 \wedge...\wedge\widehat{\mathrm{d}u_j}\wedge...\wedge \mathrm{d}u_r .$$

LEMMA I.3.2.– *Assume* $1 \leq \nu \leq n-1$. *Let* ψ , u_1 ,..., u_n *be* $n+1$ C^∞ *functions in* $\Omega\times\partial\Omega\times[0,1]$. *We have*

(I.3.12)
$$\sum_{|J|=\nu+1} \sum_{j\in J} \epsilon(j(J\backslash j))\psi u_j \mathbb{D}\, a_{J\backslash j}\wedge \mathbb{D}(\psi u)_{(J)} =$$

$$\psi^{n-\nu} \sum_{|J|=\nu+1} \sum_{j\in J} \epsilon(j(J\backslash j)) u_j \mathbb{D}\, a_{J\backslash j}\wedge \mathbb{D}u_{(J)} .$$

Proof: Consider the sum

$$\sum_{|J|=\nu+1} \sum_{j\in J} \epsilon(j(J\backslash j))\psi u_j \mathbb{D}\, a_{J\backslash j}\wedge \mathbb{D}(\psi u)_{(J)} =$$

$$\psi^{n-\nu} \sum_{|J|=\nu+1} \sum_{j\in J} \epsilon(j(J\backslash j)) u_j \mathbb{D}\, a_{J\backslash j}\wedge \mathbb{D}u_{(J)} +$$

$$(-1)^{\nu}\psi^{n-\nu-1}\mathrm{d}\psi\wedge\sum_{|J|=\nu+1}\sum_{j\in J}\sum_{k\notin J}\epsilon(j(J\backslash j))\epsilon(kJ)u_{j}u_{k}\mathbb{D}\,a_{J\backslash j}\wedge(\mathbb{D}u)_{(kJ)}\;.$$

To each triplet (J,j,k) with $j\in J$, $k\notin J$, we can assign the triplet (K,k,j) with K $= [k(J\backslash j)]$. Note that $J\backslash j = K\backslash k$ and that

$$\epsilon(j(J\backslash j))\epsilon(kJ) = -\,\epsilon(k(K\backslash k))\epsilon(jK)\;,$$

which implies

$$\sum_{|J|=\nu+1}\sum_{j\in J}\sum_{k\notin J}\epsilon(j(J\backslash j))\epsilon(kJ)u_{j}u_{k}\mathbb{D}\,a_{J\backslash j}\wedge(\mathbb{D}u)_{(kJ)}\equiv 0\;,$$

whence our assertion. ☻

PROPOSITION I.3.1.– *The pull–back of the form* $F_{a}^{(\nu)}$ *to the subspace* $t = 0$ *of the space* $\Omega\times\partial\Omega\times[0,1]$ *is equal to that of* $E_{a}^{(\nu)}$ *(see (I.2.8)).*

Proof: Apply (I.3.11) to $\psi = 1/|\lambda|^{2}$, $u_{j} = \overline{\lambda}_{j}$; the assertion in Lemma I.3.1 becomes evident when $\nu = 0$. Assume $1 \leq \nu \leq n - 1$. Applying (I.3.12) to $\psi = 1/|\lambda|^{2}$ and $u_{j} = \overline{\lambda}_{j}$ yields

$$F_{a}^{(\nu)} = (-1)^{\nu}(n-\nu-1)!|\lambda|^{2(\nu-n)}\sum_{|J|=\nu+1}\sum_{j\in J}\epsilon(j(J\backslash j))\overline{\lambda}_{j}\mathbb{D}\,a_{J\backslash j}\wedge D\overline{\lambda}_{(J)}\wedge D\lambda\;,$$

which entails what we wanted if we recall the definitions of $\mathscr{E}_{j}^{(\nu)}$ and $E_{a}^{(\nu)}$, (I.2.2), (I.2.8). ☻

COROLLARY I.3.1.– *The pull–back of the form* F_{a} *to the subspace* $t = 0$ *of the space* $\Omega\times\partial\Omega\times[0,1]$ *is equal to that of* E_{a} .

PROPOSITION I.3.2.– *Assume* $0 \leq \nu \leq n - 1$. *The pull-back of the form* $F_a^{(\nu)}$ *to the subspace* $t = 1$ *of the space* $\Omega \times \partial\Omega \times [0,1]$ *is equal to*

$$(I.3.13) \qquad S_a^{(\nu)}(z,z') = (-1)^\nu (n-\nu-1)! [\partial\rho(z') \cdot \lambda]^{\nu-n} \sum_{|J|=\nu+1} \sum_{j \in J} \epsilon(j(J \setminus j)).$$

$$[\partial\rho(z')] D_a (z,z',1)_{\wedge j} \wedge D[\partial\rho(z')]_{(J)} \wedge D\lambda .$$

Proof : It suffices to apply (I.3.12) to $\psi = [\partial\rho(z') \cdot \lambda]^{-1}$ and $u_j = \partial\rho(z')$ ($j = 1,...,n$). ☻

COROLLARY I.3.2.– *The pull-back of the form* F_a *to the subspace* $t = 1$ *of the space* $\Omega \times \partial\Omega \times [0,1]$ *is equal to*

$$(I.3.14) \qquad S_a = \sum_{\nu=0}^{n-1} (-1)^{n(n+1)/2} S_a^{(\nu)} .$$

In passing, notice that

$$(I.3.15) \qquad S_a^{(0)}(z,z') = n! [\partial\rho(z') \cdot \lambda]^{-n} D[\partial\rho(z')] \wedge D\lambda .$$

We now define the following operators

$$(I.3.16) \quad B_{\partial\Omega}^a f(z) = (-2\imath\pi)^{-n} \int_{\partial\Omega \times [0,1]} e^{a(z,z',t) \cdot (z-z')} f(z') \wedge F_a(z,z',t) ,$$

$$(I.3.17) \quad \mathcal{R}_{\partial\Omega}^a f(z) = (-2\imath\pi)^{-n} \int_{\partial\Omega \times [0,1]} e^{a(z,z',t) \cdot (z-z')} f(z') \wedge D_a(z,z',t) \wedge D\lambda ,$$

$$(I.3.18) \qquad S_{\partial\Omega}^{a_1} f(z) = (-2\imath\pi)^{-n} \int_{\partial\Omega} e^{a(z,z',1) \cdot (z-z')} f(z') \wedge S_a(z,z') .$$

As before f is an arbitrary element of $C^\infty(\mathscr{C}\ell\,\Omega;\Lambda^{n,q})$; z belongs to Ω . We must state precisely our choice of orientation on $\partial\Omega\times[0,1]$. The positive volume form on $\partial\Omega$ shall be the pull-back τ_0 to $\partial\Omega$ of a $(2n-1)$-form τ (in some tubular neighborhood of $\partial\Omega$ in $\mathbb{C}^n$) such that $\mathrm{d}\bar{z}\wedge\mathrm{d}z/(2\imath)^n = \tau\wedge\mathrm{d}s$ where s is the arc length on the normal to $\partial\Omega$ oriented outwardly, relative to Ω . We shall take $\tau_0\wedge\mathrm{d}t$ as the positive volume form on $\partial\Omega\times[0,1]$. Because of (I.3.1) the kernels F_a and S_a are C^∞ in $\Omega\times\partial\Omega$ and thus the integrals (I.3.16) & (I.3.18) define C^∞ forms in Ω ; (I.3.16) defines a form that belongs to $C^\infty(\mathscr{C}\ell\,\Omega;\Lambda^{n,q})$. The notation $K_\Omega^{a_0}$ will have the meaning (I.2.20) with $a_0 = a\,(z,z',0)$ substituted for a ; $K_{\partial\Omega}^{a_0}$ will have the analogous meaning, with $\partial\Omega$ substituted for Ω .

THEOREM I.3.2.– *Suppose $1 \leq q \leq n-1$. For any $f \in C^\infty(\mathscr{C}\ell\,\Omega;\Lambda^{n,q})$ we have, in the open set Ω ,*

(I.3.19)

$$- (-1)^q K_{\partial\Omega}^{a_0} f = \mathrm{d}[(-1)^q B_{\partial\Omega}^a f] + (-1)^{q+1} B_{\partial\Omega}^a \mathrm{d}f + R_{\partial\Omega}^a f - (-1)^q S_{\partial\Omega}^{a_1} f \ .$$

Proof: We have

$$(-2\imath\pi)^n \mathrm{d}B_{\partial\Omega}^a f = \int_{\partial\Omega\times[0,1]} \mathbb{D}\Big\{ e^{a\,(z,z',t)\cdot(z-z')} f\,(z')\wedge F_a\,(z,z',t) \Big\} -$$

$$\int_{\partial\Omega\times[0,1]} (\mathrm{d}'+\mathrm{d}_t)\Big\{ e^{a\,(z,z',t)\cdot(z-z')} f\,(z')\wedge F_a\,(z,z',t) \Big\} =$$

$$\int_{\partial\Omega\times[0,1]} e^{a\,(z,z',t)\cdot(z-z')} \mathrm{d}f\,(z')\wedge F_a\,(z,z',t) +$$

$$(-1)^{n+q}\int_{\partial\Omega\times[0,1]} e^{a\,(z,z',t)\cdot(z-z')} f\,(z')\wedge \mathbb{D}^a F_a\,(z,z',t) -$$

$$\int_{\partial\Omega\times[0,1]} \mathrm{d}_t\Big\{ e^{a\,(z,z',t)\cdot(z-z')} f\,(z')\wedge F_a\,(z,z',t) \Big\} \ .$$

But, because our choice of orientation,

$$\int_{\partial\Omega\times[0,1]} d_t\left\{ e^{\boldsymbol{a}\,(z,z',t)\cdot(z-z')}f\,(z')\wedge F_{\boldsymbol{a}}\,(z,z',t)\right\} =$$

$$\int_{\partial\Omega\times\{0\}} e^{\boldsymbol{a}\,(z,z',t)\cdot(z-z')}f\,(z')\wedge F_{\boldsymbol{a}}\,(z,z',t) -$$

$$\int_{\partial\Omega\times\{1\}} e^{\boldsymbol{a}\,(z,z',t)\cdot(z-z')}f\,(z')\wedge F_{\boldsymbol{a}}\,(z,z',t) =$$

$$\int_{\partial\Omega} e^{\boldsymbol{a}\,(z,z',0)\cdot(z-z')}f\,(z')\wedge E_{\boldsymbol{a}}\,(z,z',0) - \int_{\partial\Omega} e^{\boldsymbol{a}\,(z,z',1)\cdot(z-z')}f\,(z')\wedge S_{\boldsymbol{a}}\,(z,z') \ ,$$

by virtue of Corollaries I.3.1, I.3.2. By Th. I.3.1 we get

$$(-2\iota\pi)^n\left[dB^{\boldsymbol{a}}_{\partial\Omega}f - B^{\boldsymbol{a}}_{\partial\Omega}df + (-1)^q\mathcal{R}^{\boldsymbol{a}}_{\partial\Omega}f \right] =$$

$$\int_{\partial\Omega} e^{\boldsymbol{a}\,(z,z',1)\cdot(z-z')}f\,(z')\wedge S_{\boldsymbol{a}}\,(z,z') - \int_{\partial\Omega} e^{\boldsymbol{a}\,(z,z',0)\cdot(z-z')}f\,(z')\wedge E_{\boldsymbol{a}}\,(z,z',0) \ ,$$

whence the result, by (I.2.20) and the subsequent remark, and by (I.3.18).

By combining Theorems I.2.2 & I.3.2 we obtain

THEOREM I.3.3.– *Suppose* $1 \leq q \leq n-1$. *For any* $f \in C^\infty(\mathcal{C}\ell\,\Omega;\Lambda^{n,q})$ *we have, in the open set* Ω ,

$$(I.3.20)\qquad d[(-1)^q(K^{\boldsymbol{a}0}_{\Omega} - B^{\boldsymbol{a}}_{\partial\Omega})f] + (-1)^{q+1}(K^{\boldsymbol{a}0}_{\Omega} - B^{\boldsymbol{a}}_{\partial\Omega})df =$$

$$f - \mathcal{R}^{\boldsymbol{a}0}_{\Omega}f + \mathcal{R}^{\boldsymbol{a}}_{\partial\Omega}f - (-1)^q S^{\boldsymbol{a}1}_{\partial\Omega}f \ .$$

The operator $\mathcal{R}^{\boldsymbol{a}0}_{\Omega}$ is defined as in (I.2.21) except that $\boldsymbol{a}\,(z,z')$ is to be replaced by $\boldsymbol{a}\,(z,z',0)$. Below we use the notation $\mathcal{R}^{\boldsymbol{a}1}_{\Omega}$ to mean the same operator with $a(z,z')$ replaced by $a(z,z',1)$.

There is a variant of Th. I.3.3 that will be important in later applications. We observe that

$$(-2\iota\pi)^n \mathbf{L}^a_{\partial\Omega} f(z) =$$

$$\int_{\partial(\Omega\times[0,1])} e^{a(z,z',t)\cdot(z-z')} f(z')\wedge\mathbb{D}\,a(z,z',t)\wedge D\lambda \, -$$

$$\int_{\Omega} e^{a(z,z',1)\cdot(z-z')} f(z')\wedge D\,a(z,z',1)\wedge D\lambda \, +$$

$$\int_{\Omega} e^{a(z,z',0)\cdot(z-z')} f(z')\wedge D\,a(z,z',0)\wedge D\lambda \, =$$

$$\int_{\Omega\times[0,1]} (d'+d_t)\Big[e^{a(z,z',t)\cdot(z-z')} f(z')\wedge\mathbb{D}\,a(z,z',t)\Big]\wedge D\lambda \, +$$

$$(-2\iota\pi)^n(\mathbf{L}^{a_0}_{\Omega}f - \mathbf{L}^{a_1}_{\Omega}f) \,,$$

by Stokes' theorem. [Recall that $\mathbb{D} = D + d_t$, $D = d + d'$.] But

$$\int_{\Omega\times[0,1]} (d'+d_t)\Big[e^{a(z,z',t)\cdot(z-z')} f(z')\wedge\mathbb{D}\,a(z,z',t)\Big]\wedge D\lambda \, =$$

$$\int_{\Omega\times[0,1]} \mathbb{D}\Big[e^{a(z,z',t)\cdot(z-z')} f(z')\wedge\mathbb{D}\,a(z,z',t)\Big]\wedge D\lambda \, +$$

$$d\int_{\Omega\times[0,1]} e^{a(z,z',t)\cdot(z-z')} f(z')\wedge\mathbb{D}\,a(z,z',t)\wedge D\lambda \, =$$

$$\int_{\Omega\times[0,1]} e^{a(z,z',t)\cdot(z-z')} df(z')\wedge\mathbb{D}\,a(z,z',t)\wedge D\lambda \, +$$

$$d\int_{\Omega\times[0,1]} e^{a(z,z',t)\cdot(z-z')} f(z')\wedge\mathbb{D}\,a(z,z',t)\wedge D\lambda$$

since $\Omega \times [0,1]$ is a $(2n+1)$–chain. It is natural to define:

$$(I.3.21) \quad T_\Omega^a \, f(z) = (-2\imath\pi)^{-n} \int_{\Omega \times [0,1]} e^{a(z,z',t)\cdot(z-z')} \, f(z') \wedge \mathbb{D}\,a\,(z,z',t) \wedge D\lambda \ .$$

We have proved:

LEMMA I.3.3.– *Suppose* $1 \leq q \leq n-1$. *For any* $f \in C^\infty(\mathscr{C}\diagdown\Omega;\Lambda^{n,q})$ *we have, in the open set* Ω ,

$$(I.3.22) \qquad \qquad \mathcal{R}_{\partial\Omega}^a f = \mathcal{R}_\Omega^{\infty} f - \mathcal{R}_\Omega^a f + T_\Omega^a \, df + dT_\Omega^a \, f \ .$$

Combining Th. I.3.3 & Lemma I.3.3 yields

THEOREM I.3.4.– *Suppose* $1 \leq q \leq n-1$. *For any* $f \in C^\infty(\mathscr{C}\diagdown\Omega;\Lambda^{n,q})$ *we have, in the open set* Ω ,

$$(I.3.22)$$

$$f = d\left[(-1)^q(\mathcal{K}_\Omega^{\infty} - \mathcal{B}_{\partial\Omega}^a) + T_\Omega^a\right]f + \left[(-1)^{q+1}(\mathcal{K}_\Omega^{\infty} - \mathcal{B}_{\partial\Omega}^a) + T_\Omega^a\right]df +$$

$$\mathcal{R}_\Omega^a f \ + (-1)^q S_{\partial\Omega}^a f \ .$$

I.4. VANISHING OF THE ERROR TERMS

We seek conditions ensuring that the operators $\mathcal{R}_\Omega^a$, $\mathcal{R}_{\partial\Omega}^a$, $\mathcal{S}_{\partial\Omega}^{a_1}$ in the formulas of Sections I.2 & I.3, vanish when acting on differential forms $f \in C^\infty(\mathscr{C}\angle\Omega; \Lambda^{n,q})$. For reasons that will become clearer in Ch. II it is also important to have conditions that ensure the vanishing of $\mathcal{T}_\Omega^a$.

In dealing with the results of Sect. I.2 we may assume that the map a is independent of $t \in [0,1]$. In order that $\mathcal{R}_\Omega^a f \equiv 0$ whatever $f \in C^\infty(\mathscr{C}\angle\Omega; \Lambda^{n,q})$ it is necessary and sufficient that, whatever the function $u \in C_c^\infty(\Omega)$,

$$(I.4.1) \qquad \int_\Omega u(z') \mathrm{d}\bar{z}'_I \wedge \left\{ \bigwedge_{j=1}^n (\bar{\partial} + \bar{\partial}') a_j \right\} \wedge \mathrm{d}z' \equiv 0 \ , \ \forall \ I \ , \ |I| = q \ ,$$

We have denoted by $\bar{\partial}'$ the Cauchy–Riemann operator acting in z'–space; (I.4.1) will be true if, and only if we have, when $t = 0$ and for all z, $z' \in \Omega$,

$$(I.4.2) \qquad \left\{ \bigwedge_{j\in J} \bar{\partial} a_j \right\} \wedge \left\{ \bigwedge_{k\notin J} \bar{\partial}' a_k \right\} \equiv 0 \ , \ \forall \ J \ , \ |J| = q \ .$$

More explicitly, (I.4.2) means that, for each multi–index J, $|J| = q$, we have either

$$(I.4.3) \qquad \det(\partial a_j / \partial \bar{z}_k)_{j\in J, k\in K} \equiv 0 \ , \ \forall \ K \ , \ |K| = q \ ,$$

or

$$(I.4.4) \qquad \det(\partial a_j / \partial \bar{z}'_k)_{j\notin J, k\notin K} \equiv 0 \ , \ \forall \ K \ , \ |K| = q \ .$$

EXAMPLE I.4.1.– Let p be an integer, $0 \leq p \leq n$. Take $a_j = 2\bar{z}_j$ for $j = 1,...,p$, $a_j = -2\bar{z}'_j$ for $j = p+1,...,n$ if $1 \leq p \leq n-1$; and $a = 2\bar{z}$ if $p = n$, $a = -2\bar{z}'$ if p

$= 0$. Then property (I.4.3) will hold whenever $J \nsubseteq [1,...,p]$ whereas (I.4.4) will hold whenever $[1,...,p] \nsubseteq J$. Thus the only case in which (I.4.1) may not be true is when $J = [1,..., p]$, and then we must have $p = q$. ❂

When $q = n$ condition (I.4.4) is void, and (I.4.2) reduces to

(I.4.5)
$$\det(\partial a_j / \partial \bar{z}_k)_{1 \le j,k \le n} \equiv 0 .$$

We deduce from Cor. I.2.3:

THEOREM I.4.1.– *Suppose that condition* (I.4.5) *is satisfied in* $\Omega \times \Omega$. *Then, whatever the form* $f \in C^\infty(\mathscr{C}\ell\Omega;\Lambda^{n,n})$, *we have, in* Ω ,

(I.4.6)
$$f = (-1)^n \mathrm{d} \mathcal{K}_\Omega^a f .$$

When $q = 0$ condition (I.4.3) is void and (I.4.2) reduces to

(I.4.7)
$$\det(\partial a_j / \partial \bar{z}_k^!)_{1 \le j,k \le n} \equiv 0 .$$

We deduce from Cor. I.2.2:

THEOREM I.4.2.– *Suppose that condition* (I.4.7) *is satisfied in* $\Omega \times \Omega$. *Then, whatever the form* $f \in C^\infty(\mathscr{C}\ell\Omega;\Lambda^{n,0})$ *we have, in* Ω ,

(I.4.8)
$$f = \mathcal{K}_{\partial\Omega}^a f - \mathcal{K}_\Omega^a \mathrm{d}f .$$

Formula (I.4.8) can be regarded as a generalization of the *inhomogeneous Cauchy formula*. A generalization of the standard Cauchy formula obtains by assuming $\mathrm{d}f \equiv 0$.

We look now at the case $1 \leq q \leq n-1$. First we apply Th. I.2.2. Here also we may assume that a does not depend on t .

THEOREM I.4.3.– *Suppose $1 \leq q \leq n-1$. Suppose furthermore that the pull–back of f to $\partial\Omega$ vanishes identically.*

Then, if either (I.4.3) or (I.4.4) holds in $\Omega \times \Omega$, we have, in Ω ,

$$(I.4.9) \qquad f = d[(-1)^q \mathcal{K}^a_\Omega f] + (-1)^{q+1} \mathcal{K}^a_\Omega df$$

for all $f \in C^\infty(\mathcal{C}\ell\,\Omega; \Lambda^{n,q})$.

REMARK I.4.1.– Formula (I.4.9) can be generalized to unbounded domains Ω provided we require f and df to decay sufficiently fast at infinity. Under such a constraint (I.4.9) will also be valid when $\Omega = \mathbf{C}^n$, i. e., $\partial\Omega = \emptyset$. ❸

Next we seek to exploit Th. I.3.2. Here the map a depends on $t \in [0,1]$. We note that, in order that $\mathcal{R}^a_{\partial\Omega} f \equiv 0$ whatever $f \in C^\infty(\mathcal{C}\ell\,\Omega; \Lambda^{n,q})$ it is necessary and sufficient that, whatever the function $u \in C^\infty(\partial\Omega \times [0,1])$,

$$\int_{\partial\Omega \times [0,1]} u(z',t) d\bar{z}'_I \wedge \left\{ \bigwedge_{j=1}^n (\bar{\partial} + \bar{\partial}' + d_t)\, a_j \right\} \wedge dz' \equiv 0 \ , \ \forall\ I, \ |I| = q\ .$$

This will be true if, and only if, in the "cylinder" $\Omega \times \partial\Omega \times [0,1]$,

$$(I.4.10) \qquad d_t a_i \wedge \left\{ \bigwedge_{\substack{j \in J \\ j \neq i}} \bar{\partial} a_j \right\} \wedge \left\{ \bigwedge_{k \notin J} \bar{\partial}' a_k \right\} \equiv 0 \ , \ \forall\ J, \ |J| = q+1 \ , \ i \in J \ .$$

Next we take a look at the operator $\mathcal{S}^{a_1}_{\partial\Omega}$. Let us go back to the expression (I.3.13) of the "kernels" $\mathcal{S}^{(\nu)}_a$. We must consider integrals

$$\int_{\partial\Omega} u(z')\mathrm{d}\bar{z}'_I \wedge \Big\{\bigwedge_{j\in J}(\bar{\partial}+\bar{\partial}')\,a_j\Big\} \wedge \Big\{\bigwedge_{k\in K}\bar{\partial}'[\partial_k\rho(z')]\Big\}\Big\}\wedge\mathrm{d}z'\ ,$$

where $u \in C^\infty(\partial\Omega)$, $|I| = q$, $|J| = \nu$, $|K| = n{-}1{-}\nu$, and where we put $t = 1$. The integrands are all zero as long as $\nu \leq q{-}1$. This is still true when $\nu \geq q$ provided (I.4.3) holds, when $t = 1$ and for all $z \in \Omega$, $z' \in \partial\Omega$. We derive from Th. I.3.3:

THEOREM I.4.4.– *Suppose* $1 \leq q \leq n{-}1$ *and that the following hypotheses are satisfied:* (I.4.2) *in* $\Omega\times\Omega$ *when* $t = 0$; (I.4.3) *in* $\Omega\times\partial\Omega$ *when* $t = 1$; (I.4.10) *in* $\Omega\times\partial\Omega\times[0,1]$. *Then, for any differential form* $f \in C^\infty(\mathscr{C}\ell\Omega;\Lambda^{n,q})$ *we have, in the open set* Ω ,

(I.4.11) $f = \mathrm{d}[(-1)^q(K^{\infty}_{\Omega} - B^{a}_{\partial\Omega})f] + (-1)^{q+1}(K^{\infty}_{\Omega} - B^{a}_{\partial\Omega})\mathrm{d}f$.

COROLLARY I.4.1.– *Suppose* $1 \leq q \leq n{-}1$ *and that the map* a *is independent of* t . *Suppose furthermore that* (I.4.3) *holds in* $\Omega\times\Omega$. *Then* (I.4.11) *is valid in* Ω .

Finally we apply Th. I.3.4. First of all we note that, in order that $\mathcal{R}^{a}_{\Omega}f \equiv 0$ whatever $f \in C^\infty(\mathscr{C}\ell\Omega;\Lambda^{n,q})$, condition (I.4.2) must be satisfied in $\Omega\times\Omega$, at $t = 1$. But (I.4.2) is a consequence of (I.4.3) which must be true in $\Omega\times\partial\Omega$, also at $t = 1$, if we want $S^{a}_{\Omega}f$ to vanish identically. On the other hand, in order that $T^{a}_{\Omega}f \equiv 0$ we need that, whatever the function $u \in C^\infty((\mathscr{C}\ell\Omega)\times[0,1])$,

$$\int_{\Omega\times[0,1]} u(z',t)\mathrm{d}\bar{z}'_I \wedge \Big\{\bigwedge_{j=1}^{n}(\bar{\partial}+\bar{\partial}'+\mathrm{d}_t)\,a_j\Big\}\wedge\mathrm{d}z' \equiv 0\ ,\ \forall\ I,\ |I| = q\ .$$

This will be true if, and only if, in the "cylinder" $\Omega\times\Omega\times[0,1]$,

(I.4.12) $\mathrm{d}_t a_i \wedge \Big\{\bigwedge_{\substack{j\in J\\ j\neq i}}\bar{\partial}\,a_j\Big\} \wedge \Big\{\bigwedge_{k\notin J}\bar{\partial}'\,a_k\Big\} \equiv 0\ ,\ \forall\ J,\ |J| = q\ ,\ i \in J\ .$

We must underline the difference between (I.4.10) & (I.4.12): in the former the length of the multi–index J is equal to $q+1$, whereas in the latter it is equal to q. On the other hand, in Th. I.4.4, we require that (I.4.10) be valid for $z' \in \partial\Omega$, whereas in Th. I.4.5 below, (I.4.12) must be valid for all $z' \in \Omega$.

THEOREM I.4.5.– *Suppose $1 \leq q \leq n{-}1$ and that the following hypotheses are satisfied: (I.4.3) in $\Omega{\times}\Omega$ when $t = 1$; (I.4.12) in $\Omega{\times}\Omega{\times}[0,1]$. Then, for any **closed** differential form $f \in C^\infty(\mathscr{C}\ell\Omega;\Lambda^{n,q})$ we have, in the open set Ω,*

$$(I.4.13) \qquad f = d[(-1)^q(\mathcal{K}_\Omega^\infty - \mathcal{B}_{\partial\Omega}^a)f] .$$

$$\mathscr{T}he \quad case \quad q = 1$$

When $q = 1$ (I.4.2) reads

$$(I.4.14) \qquad \bar{\partial} a_j \wedge \left\{ \bigwedge_{k \neq j} \bar{\partial}' a_k \right\} \equiv 0 , \quad \forall\, j = 1,...,n ;$$

and (I.4.3),

$$(I.4.15) \qquad \bar{\partial} a_j = 0 , \quad \forall\, j = 1,...,n .$$

As for (I.4.10) it reads

$$(I.4.16) \qquad d_t a_i \wedge \bar{\partial} a_j \left\{ \bigwedge_{\substack{k \neq i \\ k \neq j}} \bar{\partial}' a_k \right\} \equiv 0 , \quad \forall\, i , j \in [1,...,n] , \; i \neq j ;$$

and (I.4.12),

$$(I.4.17) \qquad d_t a_i \wedge \left\{ \bigwedge_{k \neq i} \bar{\partial}' a_k \right\} \equiv 0 , \quad \forall\, i = 1,...,n .$$

Th. I.4.4 requires, when $q = 1$, that (I.4.14) be satisfied in $\Omega \times \Omega$ when $t = 0$, (I.4.15) be satisfied in $\Omega \times \partial\Omega$ when $t = 1$ and that (I.4.16) be satisfied in $\Omega \times \partial\Omega \times [0,1]$. Th. I.4.5 requires, when $q = 1$, that (I.4.15) be satisfied when $t = 1$, in $\Omega \times \Omega$, and that (I.4.17) be satisfied in $\Omega \times \Omega \times [0,1]$.

CHAPTER II

HOMOTOPY FORMULAS

IN THE TANGENTIAL CAUCHY–RIEMANN COMPLEX

II.1.   LOCAL DESCRIPTION OF THE TANGENTIAL CAUCHY–RIEMANN
          COMPLEX

While $\mathscr{D}$ continues to denote a domain in $\mathbf{C}^n$ the letter $\mathscr{B}$ shall stand for an open ball in $\mathbb{R}^d$ ($d \geq 1$) centered at the origin. We also take $\mathscr{D}$ to be a neighborhood of 0 . We shall systematically write

$$(\text{II.1.1}) \qquad\qquad m = n + d \ .$$

We consider d real–valued C^∞ functions in $\mathscr{D} \times \mathscr{B}$, $\varphi_k(z,s)$ ($k = 1,...,d$) and the complex–valued functions

$$(\text{II.1.2}) \qquad\qquad w_k = s_k + \imath\varphi_k(z,s) \ , \ k = 1,...,d \ .$$

We shall often say that the functions z_j and w_k ($1 \leq j \leq n$, $1 \leq k \leq d$) define a *CR structure* on $\mathscr{D} \times \mathscr{B}$. We shall always reason under the hypothesis

$$(\text{II.1.3}) \qquad\qquad \varphi_k\big|_0 = 0 \ , \ \mathrm{d}\varphi_k\big|_0 = 0 \ , \ k = 1,...,d \ .$$

Below we often write $w = (w_1,...,w_d) : \mathscr{D} \times \mathscr{B} \to \mathbf{C}^d$, $\varphi = (\varphi_1,...,\varphi_d) : \mathscr{D} \times \mathscr{B} \to \mathbb{R}^d$. We shall denote by w_s or by $\partial w/\partial s$ the Jacobian matrix $(\partial w_j/\partial s_k)_{1 \leq j,k \leq d}$; we may write $w_s = I + \imath\varphi_s$, where I is the $d \times d$ identity matrix. Hypothesis (II.1.3) allows us to assume that, in $\mathscr{D} \times \mathscr{B}$ possibly contracted about the origin, the norm of the differential $\mathrm{d}\varphi$ is as small as we wish; we shall always suppose $\|\mathrm{d}\varphi\| << 1$; it ensues, in particular, that w_s is nonsingular.

We let $\mathcal{V}$ denote the vector subbundle of the complex(ified) tangent bundle $\mathbf{C}T(\mathscr{D} \times \mathscr{B})$ whose sections annihilate all the functions z_j and w_k . Of course,

$$(\text{II.1.4}) \qquad\qquad [\mathcal{V} , \mathcal{V}] \subset \mathcal{V} \ ,$$

that is to say, the commutation bracket of two C^1 sections of $\mathcal{V}$ is a section of $\mathcal{V}$. The sections of $\mathcal{V}$ are spanned by the vector fields

$$(II.1.5) \qquad L_j = \partial/\partial \bar{z}_j - \imath \sum_{k=1}^{d} (\partial \varphi_k / \partial \bar{z}_j) M_k \, ,$$

where

$$(II.1.6) \qquad M_k = \sum_{\ell=1}^{d} \mu_{k\ell} \partial/\partial s_\ell \, , \quad k = 1,...,d \, ,$$

and $(\mu_{k\ell})_{1 \leq k, \ell \leq d} = (I + \imath \varphi_s)^{-1}$ is the inverse matrix of w_s . Notice that, for all i , j $= 1,...,n$, k , $\ell = 1,...,d$,

$$L_j z_i = L_j w_k = 0 \, , \; L_j \bar{z}_i = \delta_{ij} \, ,$$

(II.1.7)

$$M_k z_i = M_k \bar{z}_i = 0 \, , \; M_k w_\ell = \delta_{k\ell}$$

$(\delta_{ij}$, $\delta_{k\ell}$: Kronecker indices), whence

$$(II.1.8) \qquad [L_i , L_j] = [L_j , M_k] = [M_k , M_\ell] = 0 \, .$$

We denote by T' the vector subbundle of the complex cotangent bundle $\mathbb{C}T^*(\mathcal{D} \times \mathcal{B})$ spanned by the differentials dz_j and dw_k ; T' is the orthogonal of $\mathcal{V}$ for the natural duality between tangent and cotangent vectors. We have

$$(II.1.9) \qquad dT' \subset T' \, ,$$

that is to say, the exterior derivative of any C^1 section of T' belongs to the ideal generated by T' in the exterior algebra over $\mathcal{D} \times \mathcal{B}$.

We denote by $T'^{p,q}$ (p , $q \in \mathbb{Z}_+$) the vector subbundle of the $(p+q)$ exterior power of $\mathbb{C}T^*(\mathscr{D} \times \mathscr{B})$ spanned by the exterior products

$$\omega_1 \wedge \ldots \wedge \omega_{p+q}$$

in which the ω_i are differential forms, at least p of which are equal to some dz_j or some dw_k . We have, for $0 \le p \le m-1$, $1 \le q \le n$,

(II.1.10) $T'^{p+1,q-1} \subset T'^{p,q}$.

Actually we have

(II.1.11) $T'^{p+1,q-1} = T'^{p,q}$ *if* $q > n$.

Also,

(II.1.12) $T'^{m,n} =$ *the $(m+n)$ exterior power of* $\mathbb{C}T^*(\mathscr{D} \times \mathscr{B})$;

(II.1.13) $T'^{p,q} = 0$ *if* $p > m$ *or* $p+q > m + n$.

This allows us to define the quotient vector bundles

(II.1.14) $\Lambda^{p,q} = T'^{p,q}/T'^{p+1,q-1}$.

We agree that

(II.1.15) $\Lambda^{p,0} = T'^{p,0}$;

$\Lambda^{0,0}$ is the trivial bundle $(\mathscr{D} \times \mathscr{B}) \times \mathbb{C}$. Moreover,

(II.1.16) $$\Lambda^{m,q} = T'^{m,q} \; ;$$

(II.1.17) $$\Lambda^{p,q} = 0 \;\; if \;\; p > m \;\; or \;\; q > n \;.$$

It is readily checked that there are natural vector bundle isomorphisms

(II.1.18) $$\Lambda^{p,q} \cong \Lambda^{p,0} \otimes \Lambda^{0,q} \;,$$

(II.1.19) $$\Lambda^{0,q} \cong \Lambda^{m,q} \;,$$

so that

(II.1.20) $$\Lambda^{p,q} \cong \Lambda^{p,0} \otimes \Lambda^{m,q} \;.$$

The isomorphism (II.1.19) is defined by the map $\omega \to \omega \wedge dz \wedge dw$. Here, as troughout the sequel,

(II.1.21) $$dz = dz_1 \wedge ... \wedge dz_n \;,\; dw = dw_1 \wedge ... \wedge dw_d \;.$$

Next we avail ourselves of the following consequence of (II.1.9):

(II.1.22) $$dT'^{p,q} \subset T'^{p,q+1}.$$

Let $\mathcal{O}$ be any open subset of $\mathcal{D} \times \mathcal{B}$. According to (II.1.22) the exterior derivative induces differential operators

(II.1.23) $$\vartheta^{p,q} : \mathcal{D}'(\mathcal{O};\Lambda^{p,q}) \to \mathcal{D}'(\mathcal{O};\Lambda^{p,q+1}) \;,\; q = 0,1,...,$$

(II.1.24) $$\vartheta^{p,q} : C^\infty(\mathcal{O};\Lambda^{p,q}) \to C^\infty(\mathcal{O};\Lambda^{p,q+1}) \;,\; q = 0,1,....$$

[As a general rule, if E is a complex vector bundle over a C^∞ manifold $\mathcal{M}$ we denote by $C^\infty(\mathcal{M};E)$ the space of C^∞ sections of E over $\mathcal{M}$, and by $\mathcal{D}'(\mathcal{M};E)$ the space of distribution sections.] If the closure of $\mathcal{O}$ is compact we also have

$$(\text{II}.1.25) \qquad \vartheta^{p,q} C^\infty(\mathscr{C}\!\angle\mathcal{O};\Lambda^{p,q}) \subset C^\infty(\mathscr{C}\!\angle\mathcal{O};\Lambda^{p,q+1}).$$

The sequences (II.1.23) & (II.1.24) define differential complexes since obviously we have $\vartheta^{p,q+1} \circ \vartheta^{p,q} = 0$. We refer to (II.1.24) as the *Cauchy–Riemann* (abbreviated to *CR*) *differential complex* in $\mathcal{O}$.

We shall focus on the case $p = m$. The sections of $\Lambda^{m,q}$ over $\mathcal{O}$ can be identified to true differential forms, in contrast to what happens when $p < m$, in which case they are equivalence classes of differential forms. An element of $C^\infty(\mathcal{O};\Lambda^{m,q})$ will be a form

$$(\text{II}.1.26) \qquad f = \sum_{|J|=q} f_J(z,s)\mathrm{d}\bar{z}_J \wedge \mathrm{d}z \wedge \mathrm{d}w ,$$

with $f_J \in C^\infty(\mathcal{O})$ if $|J| = q$. The operator $\vartheta^{m,q}$ can be identified to the exterior derivative:

$$(\text{II}.1.27) \qquad \mathrm{d}f = \sum_{|J|=q} \sum_{k=1}^{n} L_k f_J(z,s)\mathrm{d}\bar{z}_k \wedge \mathrm{d}\bar{z}_J \wedge \mathrm{d}z \wedge \mathrm{d}w .$$

We shall use standard terminology: if $\mathrm{d}f = 0$ we say that f is a *cocycle*, or that f is *closed*. If $q \geq 1$ and if $f = \mathrm{d}u$ for some C^∞ section u of $\Lambda^{m,q-1}$ we say that f is a *coboundary* or that f is *exact*. We shall use the same terminology when the coefficients are distributions, *i. e.*, when $f \in \mathcal{D}'(\mathcal{O};\Lambda^{m,q})$. When f is smooth one might want to be a bit more precise and state whether u is smooth or not.

If we go back to the expressions (II.1.5) we see that

$$(\text{II}.1.27) \qquad \mathrm{d}f \;=\; \bar{\partial}_z f \;-\; \imath \sum_{k=1}^{d} \bar{\partial}_z \varphi_k \wedge \mathrm{M}_k f \,,$$

where the vector fields M_k act on the form f coefficientwise:

$$(\text{II}.1.29) \qquad \mathrm{M}_k f \;=\; \mathrm{d}z \wedge \mathrm{d}w \wedge \sum_{|J|=q} \mathrm{M}_k f_J \, \mathrm{d}\bar{z}_J \,.$$

The action of $\bar{\partial}_z$ is the standard one (see (I.1.2)).

We are also going to need the concepts of characteristic set and Levi form.

By definition the *characteristic set* of the structure is the intersection with the real cotangent bundle $T^*(\mathscr{D} \times \mathscr{B})$ of the cotangent structure bundle T' (whose fibres are spanned by the differentials $\mathrm{d}z_i$ and $\mathrm{d}w_k$); we shall denote it by T^0. Since T' is the orthogonal of $\mathcal{V}$, T^0 is the set of common zeros in $T^*(\mathscr{D} \times \mathscr{B})$ of all the symbols of the sections of the tangent structure bundle $\mathcal{V}$. Note that $T' \cap \overline{T'} = T^0 \otimes \mathbb{C}$ (the fibres of $\overline{T'}$ are spanned by the $\mathrm{d}\bar{z}_i$ and $\mathrm{d}\bar{w}_k$). Since obviously $T' + \overline{T'} = \mathbb{C}T^*(\mathscr{D} \times \mathscr{B})$ we conclude that $T' \cap \overline{T'}$ is a vector subbundle of $\mathbb{C}T^*(\mathscr{D} \times \mathscr{B})$. A consequence of (II. 1.3) is that, at the origin, $\mathrm{d}w_k = \mathrm{d}\bar{w}_k = \mathrm{d}s_k$. Since the $\mathrm{d}z_i$ and $\mathrm{d}\bar{z}_j$ are linearly independent we conclude that

(II.1.30)   *the fibre of* T^0 *at the origin is the* $\mathbb{R}$*-linear span of* $\mathrm{d}s_1$,..., $\mathrm{d}s_d$.

Thus *the integer d is equal to the rank of the characteristic bundle T^0.*

Let now $\wp = (z_0, s_0)$ be an arbitrary point of $\mathscr{D} \times \mathscr{B}$, $\mathbf{v}$ an element of the fibre of $\mathcal{V}$ at $\wp$, ϖ an element of the fibre of T^0 at $\wp$. Select at random a smooth section L of $\mathcal{V}$ in a neighborhood of $\wp$ whose value at $\wp$ is equal to $\mathbf{v}$. It is well known, and readily checked, that the number

$$(\text{II}.1.31) \qquad < \varpi \,, [L, \bar{L}]/2\imath \big|_{\wp} >$$

is independent of the choice of the section L . We shall denote it by $\mathcal{L}_{\wp,\varpi}(\mathbf{v})$; $\mathcal{L}_{\wp,\varpi}$ is a hermitian quadratic form on the complex vector space $V_{\wp}$. It is called the *Levi form of the CR structure at the characteristic point* $(\wp,\varpi) \in T^0$. We shall express it in the basis $L_1,...,L_n$, that is to say, represent it by the self–adjoint matrix

(II.1.32) $$(< \varpi , [L_i , \bar{L}_j]/2\imath \big|_{\wp} >)_{i,j=1,...,n} ,$$

to which we refer as the *Levi matrix.*

Take $\wp$ to be the origin. According to what was said earlier we have

$$\varpi = \sigma \cdot ds = \sigma_1 ds_1 +...+ \sigma_d ds_d \quad for \ some \ \sigma \in \mathbb{R}^d.$$

On the other had, it follows from (II.1.3), (II.1.5) & (II.1.6) that

$$L_i\big|_0 = \partial/\partial\bar{z}_i \quad , \quad M_k\big|_0 = \partial/\partial s_k ,$$

whence, thanks once again to (II.1.3),

(II.1.33)   *at the origin* ,

$$[L_i , \bar{L}_j]/2\imath = \sum_{k=1}^{d}(\partial^2\varphi_k/\partial\bar{z}_i\partial z_j)\partial/\partial s_k ,$$

and consequently,

(II.1.34)   *at the origin, the Levi matrix is equal to the complex Hessian*

$$\sigma \cdot \varphi_{z\bar{z}} = (\partial^2(\sigma\cdot\varphi)/\partial\bar{z}_i\partial z_j)_{i,j=1,...,n} \ .$$

One might also associate to the CR structure various *Levi maps*. Thus for instance, we can associate at the origin the quadratic map

$$\mathbb{C}^n \ni \zeta \;\to\; \varphi_{z\bar{z}} \zeta \cdot \bar{\zeta} \in \mathbb{R}^d \;.$$

More generally, we get a map from the fibre of $\mathcal{V}$ at $\wp$ into the fibre of (the dual of) the characteristic set T^0 at the same point. Etc.

There is an alternative, often used, representation of the CR structure we are considering. The map $(z,s) \to (z,w)$ with w defined by (II.1.2) is a diffeomorphism of $\mathcal{D} \times \mathcal{B}$ onto a C^∞ submanifold Σ of $\mathbb{C}^m$. The latter is defined by the equations

(II.1.35) $\mathscr{I}m \, w_k - \varphi_k(z, \mathscr{R}e \, w) = 0 \; , \; k = 1,...,d \;.$

Thus $\mathrm{codim}_{\mathbb{R}} \, \Sigma = d$, $\dim_{\mathbb{R}} \Sigma = m+n$. The case $d = 1$ is particularly important: Σ is a real hypersurface in $\mathbb{C}^{n+1}$. In passing note that this is also the case where the characteristic set has rank one; the complement of the zero section in T^0 consists of two connected "half–planes", each corresponding to one side of Σ . When $d = 1$ one often says that the CR structure on $\mathcal{D} \times \mathcal{B}$ is of the *hypersurface type*.

Back to the general case $d \geq 1$ the geometric meaning of (II.1.3) is clear: the submanifold Σ passes through the origin and its tangent space at 0 is the linear subspace $\mathscr{I}m \, w = 0$ (its holomorphic tangent space is the complex linear subspace $w = 0$). Furthermore the submanifold Σ is *generic*; this means that the pull–backs to Σ of the differentials dz_j , dw_k in $\mathbb{C}^m$ $(1 \leq j \leq n$, $1 \leq k \leq d)$ are $\mathbb{C}$–linearly independent. Another manner of stating the same fact is by saying that, if we call ρ_k the left sides in the equations (II.1.35), then we have

(II.1.36) $\partial\rho_1 \wedge ... \wedge \partial\rho_d \neq 0$

at every point of Σ .

The **push–forward**, under the map $(z,s) \to (z,w)$, of the vector bundle $\mathcal{V}$ is the *tangential Cauchy–Riemann vector bundle* on Σ : its sections are those linear combinations of the vector fields $\partial/\partial\bar{z}_j$ and $\partial/\partial\bar{w}_k$ that are tangent to Σ . It is sometimes denoted by $T_\Sigma^{0,1}$.

Let $T'^{1,0}$ denote the vector subbundle of $\mathbb{C}T^*(\mathbb{C}^m)$ whose sections are linear combinations of the dz_j and dw_k . We shall denote by $T_\Sigma'^{1,0}$ the pull–back of $T'^{1,0}$ to Σ ; it is the orthogonal of $T_\Sigma^{0,1}$. The pull–back of $T_\Sigma'^{1,0}$ to $\mathcal{D}\times\mathcal{B}$ under the map $(z,s) \to (z,w)$ is evidently equal to T'.

If we identify the open set $\mathcal{O} \subset \mathcal{D}\times\mathcal{B}$ to its image via the map $(z,s) \to (z,w)$ and thus regard it as an open subset of Σ the differential complexes (II.1.24) & (II.1.25) become the *tangential Cauchy–Riemann complexes* on Σ . In this interpretation we write $\bar{\partial}_\Sigma$ rather than $\vartheta^{p,q}$.

It is also clear that any differential form (or any current) (II.1.26) can be regarded as a form (or a current) on Σ , *i. e.*, it can be identified to its push–forward under the diffeomorphism $(z,s) \to (z,w)$.

II.2.   APPLICATION OF THE BOCHNER–MARTINELLI FORMULA
TO A CR MANIFOLD

We are going to extend the formulas of Sect. I.2 after introducing the variables $s = (s_1, ..., s_d)$, $s' = (s'_1, ..., s'_d)$ $(s , s' \in \mathcal{B})$. This will require that we handle partial differentiation and partial integrations, not only with respect to z' but also with respect to s'. Throughout the remainder of the present chapter, unless otherwise specified d will stand for the exterior derivative in (z,s)–space, d' for that in (z',s')–space; as before, $D = d + d'$.

Let q be an integer, $0 \leq q \leq n$, and let Ω denote, as in Ch. I , a relatively compact open subset of $\mathcal{D}$. We introduce a relatively compact open subset Θ of $\mathcal{B}$, also with a C^∞ boundary $\partial\Theta$; set $\mathcal{O} = \Omega \times \Theta$. The rule for partial integration with respect to (z',s') is essentially the same as that for partial integration with respect to z' described in Sect. I.1 ; the only difference is the presence of the real variables s_k and s'_ℓ . We deal with differential forms in (z,s,z',s')–space represented as sums

(II.2.1)
$$ F = \sum_{|J|=q} F_J \wedge \mathrm{d}\bar{z}_J \wedge \mathrm{d}z \wedge \mathrm{d}w , $$

where $\mathrm{d}s = \mathrm{d}s_1 \wedge ... \wedge \mathrm{d}s_d$ and the F_J are differential forms in (z',s')–space $\mathcal{O}$ whose coefficients are functions of (z,s,z',s') endowed with a reasonable degree of regularity. We shall only consider forms such that the total degree of every F_J is either equal to $m+n$ $(= 2n+d)$ or to $m+n-1$. The definition of the "partial integral" with respect to (z',s') is obvious:

(II.2.2)
$$ \int_{\mathcal{O}} F = \sum_{|J|=q} \left\{ \int_{\mathcal{O}} F_J \right\} \mathrm{d}\bar{z}_J \wedge \mathrm{d}z \wedge \mathrm{d}w . $$

or the analogous formula with $\partial\mathcal{O}$ substituted for $\mathcal{O}$. Notice that,

$$dF = (-1)^r \sum_{|J|=q} \sum_{k=1}^{n} L_k F_J \wedge d\bar{z}_k \wedge d\bar{z}_J \wedge dz \wedge dw , \quad d'F = \sum_{|J|=q} d'F_J \wedge d\bar{z}_J \wedge dz \wedge dw ,$$

where L_k acts coefficientwise on F_J and $r = \deg F_J$. As a consequence of this and of the Stokes' theorem we have

$$(\text{II.2.3}) \quad d\int_O F = (-1)^d \int_O dF , \quad d\int_{\partial O} F = (-1)^{d-1} \int_{\partial O} dF , \quad \int_{\partial O} F = \int_O d'F .$$

The map a in (I.1.8) will be allowed to depend on s , s'. Thus we are now dealing with a C^∞ map

$$a : (\mathscr{D} \times \mathscr{B}) \times (\mathscr{D} \times \mathscr{B}) \to \mathbb{C}^n.$$

Formula (I.1.10) and the notation (I.1.11) remain valid.

We also extend the definitions (I.2.8), (I.2.9); one must keep, in mind, however that now a depends on $(z,s,z',s') \in (\mathscr{D} \times \mathscr{B}) \times (\mathscr{D} \times \mathscr{B})$ and that D stands for the total differential in (z,s,z',s')–space. Thus $E_a = E_a(z,s,z',s')$. With this interpretation the identity (I.2.13) remains valid.

Let f be given by (II.1.26). We define (cf. (I.2.20), (I.2.21)):

$$(\text{II.2.4}) \qquad K_O^a f(z,s) = (-2\imath\pi)^{-n} \int_O e^{a \cdot (z-z')} f(z',s') \wedge E_a \wedge dw ;$$

$$(\text{II.2.5}) \qquad \mathcal{R}_O^a f(z,s) = (-2\imath\pi)^{-n} \int_O e^{a \cdot (z-z')} f(z',s') \wedge D\, a \wedge dz \wedge dw ,$$

where it is understood that a and E_a depend on (z,s,z',s') and where we are using partial integration with respect to (z',s'). We shall also use the notation $K_{\partial O}^a f$: its definition differs from (II.2.4) only in that the integration with respect to (z',s') is carried out over the boundary ∂O of O . Note that $K_O^a f \equiv 0$ when $q = 0$, by the definition of E_a, while $K_{\partial O}^a f \equiv 0$ when $q = n$, since f is an $(m+n)$-form.

LEMMA II.2.1.– *If* $0 \leq q \leq n$ *, for any* $f \in C^\infty(\mathcal{B} \angle \mathcal{O}; \Lambda^{m,q})$ *we have, in the open set* $\mathcal{O}$*,*

(II.2.6) $\quad (2\pi)^{-n} \int_{\mathcal{O}} e^{a \cdot (z-z')} f(z',s') \wedge D^a E_a \wedge dw = \int_\Theta f(z,s') \wedge dw - \mathcal{R}_\mathcal{O}^a f(z,s)$.

Proof: There is a differential form F of bidegree (n,q) in z–space Ω , whose coefficients are C^∞ functions in $\mathcal{B} \angle \mathcal{O}$, such that $f(z,s) \equiv F(z,s) \wedge ds$ modulo exterior products in which at least one factor ds_ℓ is missing. Clearly,

$$\int_\Theta f(z,s') = \int_\Theta F(z,s') \wedge ds'$$

where we use partial integration with respect to s'. By (I.2.22) we know that

$$(2\pi)^{-n} \int_\Omega e^{a \cdot (z-z')} F(z',s') \wedge D^a E_a = F(z,s') - \mathcal{R}_\Omega^a F(z,s') .$$

We form the exterior product of both sides of this equation with ds', from the right, and we carry out the integration with respect to s', over Θ :

$$(2\pi)^{-n} \int_{\mathcal{O}} e^{a \cdot (z-z')} F(z',s') \wedge ds' \wedge D^a E_a = \int_\Theta f(z,s') -$$

$$(2\pi)^{-n} \int_{\mathcal{O}} e^{a \cdot (z-z')} f(z',s') \wedge D a \wedge dz ,$$

whence (II.2.6). ∎

THEOREM II.2.1.– *Suppose* $0 \leq q \leq n$ *. For any* $f \in C^\infty(\mathcal{B} \angle \mathcal{O}; \Lambda^{m,q})$ *we have, in the open set* $\mathcal{O}$ *,*

(II.2.7)

$$\int_\Theta f(z,s') \wedge dw = d[(-1)^q \mathcal{K}_\mathcal{O}^a f] + (-1)^{d+q+1} \mathcal{K}_\mathcal{O}^a df + (-1)^{d+q} \mathcal{K}_{\partial\mathcal{O}}^a f + \mathcal{R}_\mathcal{O}^a f .$$

Proof: It parallels the proof of Th. I.2.2. We have, according to (II.2.3),

$$(-2\imath\pi)^n d\mathcal{K}_{\mathcal{O}}^{a} f\,(z,s)\,=\,(-1)^{n+q}\int_{\mathcal{O}} e^{a\,\cdot(z-z')} f\,(z',s')\wedge(D^{a}E_{a}-d'^{a}E_{a})\wedge dw$$

and by the Stokes theorem,

$$-\,(-1)^{n+d+q}\int_{\mathcal{O}} e^{a\,\cdot(z-z')} f\,(z',s')\wedge d'^{a}E_{a}\,=$$

$$\int_{\mathcal{O}} e^{a\,\cdot(z-z')} df\,(z',s')\wedge E_{a}\,-\,\int_{\partial\mathcal{O}} e^{a\,\cdot(z-z')} f\,(z',s')\wedge E_{a}\ ,$$

whence, by (II.2.6),

$$(-1)^q d\mathcal{K}_{\mathcal{O}}^{a} f\,-\,(-1)^{d+q}(\mathcal{K}_{\mathcal{O}}^{a} df\,-\,\mathcal{K}_{\partial\mathcal{O}}^{a} f\,)\,=\,\int_{\theta} f\,(z,s')\wedge dw\,-\,\mathcal{R}_{\mathcal{O}}^{a} f\,.\ \blacksquare$$

Next we examine under which conditions the integral operator $\mathcal{R}_{\mathcal{O}}^{a}$ acting on forms $f\in C^{\infty}(\mathscr{C}\ell\,\mathcal{O};\Lambda^{m,q})$ according to formula (II.2.5) vanishes identically. We assume the expression of f to be (II.1.26). We must look at integrals over $\mathcal{O}$ (as a subset of (z',s')–space) of forms

$$f_I(z',s')dz'\wedge dw'\wedge d\bar{z}'_I\wedge D\,a\,\wedge dz\wedge d\bar{w}\ ,\ \ |I|\,=\,q\ .$$

By analogy with the argument at the beginning of Sect. I.4 we conclude that the following differential n–form must vanish:

$$(\text{II}.2.8)\qquad\qquad \left\{\bigwedge_{j\in J} L\,a_j\right\}\wedge\left\{\bigwedge_{k\notin J} L'\,a_k\right\}\,\equiv\,0\ ,\ \ \forall\,J,\ |J|\,=\,q\ .$$

We have used the notation

$$(\text{II.2.9}) \qquad\qquad Lg \; = \; \sum_{i=1}^{n} L_i g \; d\bar{z}_i$$

and L' for the analogue in (z',s')–space (here g is a function). We observe that the validity in $\mathcal{O} \times \mathcal{O}$ of (II.2.8) is a consequence of either one of the following two properties:

$$(\text{II.2.10}) \qquad \forall \; J \; , \; K \; , \; |J| \; = \; |K| \; = \; q \; , \quad \det(L_j a_k)_{j \in J, k \in K} \equiv 0 \; ;$$

$$(\text{II.2.11}) \qquad \forall \; J \; , \; K \; , \; |J| \; = \; |K| \; = \; n{-}q \; , \quad \det(L'_j a_k)_{j \in J, k \in K} \equiv 0 \; .$$

LEMMA II.2.2.– *Let q be an integer, $0 \leq q \leq n$. Suppose that (II.2.8) holds, in $\mathcal{O} \times \mathcal{O}$. Then, whatever $f \in C^\infty(\mathcal{C} \mathcal{L} \mathcal{O}; \Lambda^{m,q})$, we have $\mathcal{K}_\mathcal{O}^a f \equiv 0$ in $\mathcal{O}$.*

THEOREM II.2.2.– *Let q be an integer, $0 \leq q \leq n$. Suppose (II.2.8) holds in $\mathcal{O} \times \mathcal{O}$. Then, for all differential forms $f \in C^\infty(\mathcal{C} \mathcal{L} \mathcal{O}; \Lambda^{m,q})$, we have, in $\mathcal{O}$,*

$$(\text{II.2.12}) \quad \int_\theta f(z,s') \wedge dw \; = \; d[(-1)^q \mathcal{K}_\mathcal{O}^a f] \; + \; (-1)^{d+q+1} \mathcal{K}_\mathcal{O}^a df \; + \; (-1)^{d+q} \mathcal{K}_{\partial \mathcal{O}}^a f \; .$$

When $q = 0$ condition (II.2.10) is void and (II.2.11) reads:

$$(\text{II.2.13}) \qquad\qquad \det(L'_j a_k)_{1 \leq j, k \leq n} \equiv 0 \; .$$

When $q = n$ condition (II.2.11) is void and (II.2.10) reads:

$$(\text{II.2.14}) \qquad\qquad \det(L_j a_k)_{1 \leq j, k \leq n} \equiv 0 \; .$$

COROLLARY II.2.1.– *Suppose (II.2.13) holds in $\mathcal{O} \times \mathcal{O}$. Then, for all differential forms $f \in C^\infty(\mathcal{C} \mathcal{L} \mathcal{O}; \Lambda^{m,0})$, we have, in $\mathcal{O}$,*

(II.2.15) $(-1)^d \int_\theta f(z,s') \wedge dw = - \kappa_\theta^a df + \kappa_{\partial\theta}^a f$.

Indeed, if f is an $(m,0)$–form, the integral with respect to (z',s') of f $(z',s') \wedge E_a$ over θ vanishes identically.

COROLLARY II.2.2.– *Suppose* (II.2.14) *holds in* $\theta \times \theta$. *Then, for all differential forms* f $\in C^\infty(\mathscr{C} \angle \theta; \Lambda^{m,n})$, *we have, in* θ ,

(II.2.16) $\int_\theta f(z,s') \wedge dw = d[(-1)^m \kappa_\theta^a f]$.

Indeed, if the form f has top–degree, $m+n$, then $df \equiv 0$ and the integral of f over $\partial\theta$, thus $\kappa_{\partial\theta}^a f$, vanishes identically.

The next step is to introduce a new set of variables $\sigma = (\sigma_1, ..., \sigma_d) \in \mathbb{R}^d$, which we regard as the variables dual to the s_k . Let Γ be an open cone in $\mathbb{R}^d \backslash \{0\}$: if $\sigma \in \Gamma$, $c > 0$, then $c\sigma \in \Gamma$. Henceforth we allow a to depend also on $\sigma \in \Gamma$. We shall make the following hypothesis:

(II.2.17)   *for every* $i = 1, ..., n$, a_i *is a measurable function of* $\sigma \in \Gamma$ *valued in the space of* C^∞ *functions of* $(z,s,z',s') \in (\mathscr{D} \times \mathscr{B}) \times (\mathscr{D} \times \mathscr{B})$;

(II.2.18] $a(z,s,z',s',\sigma)| \leq const.|\sigma|$, $\forall (z,s,z',s') \in (\mathscr{D} \times \mathscr{B}) \times (\mathscr{D} \times \mathscr{B})$, $\sigma \in \Gamma$.

On the other hand, we replace in (II.2.7), f by $e^{-i\sigma \cdot w} f$, and we multiply both sides by $e^{i\sigma \cdot w}$. Notice that this multiplication commutes with the exterior derivative, thanks to the presence of the factor dw . By Th. II.2.1 we obtain, for all (z,s) $\in \theta$ and all $\sigma \in \Gamma$,

(II.2.19) $\int_\theta e^{i\sigma \cdot [s+i\varphi(z,s)-s'-i\varphi(z,s')]} f(z,s') \wedge dw =$

$$d\left\{(-1)^q(-2\imath\pi)^{-n}\int_{\mathcal{O}}e^{\imath\sigma\cdot(w-w')+a\cdot(z-z')}f(z',s')\wedge E_a\wedge dw\right\} +$$

$$(-1)^{q+d+1}(-2\imath\pi)^{-n}\int_{\mathcal{O}}e^{\imath\sigma\cdot(w-w')+a\cdot(z-z')}df(z',s')\wedge E_a\wedge dw +$$

$$(-1)^{q+d}(-2\imath\pi)^{-n}\int_{\partial\mathcal{O}}e^{\imath\sigma\cdot(w-w')+a\cdot(z-z')}f(z',s')\wedge E_a\wedge dw +$$

$$(-2\imath\pi)^{-n}\int_{\mathcal{O}}e^{\imath\sigma\cdot(w-w')+a\cdot(z-z')}f(z',s')\wedge D_a\wedge dz\wedge dw .$$

We must keep in mind that now a and E_a depend not only on (z,s,z',s') but also on $\sigma\in\Gamma$; we have used the notation $w = s + \imath\varphi(z,s)$, $w' = s' + \imath\varphi(z',s')$.

Lastly, we multiply both sides of (II.2.19) by $(2\pi)^{-d}e^{-\varepsilon|\sigma|^2}$ ($\varepsilon > 0$) and integrate the resulting forms with respect to σ , over Γ . We introduce the notation:

(II.2.20) $\qquad\qquad\qquad I_{\theta,\Gamma}^{\varepsilon}f(z,s) =$

$$(2\pi)^{-d}\int_{\Gamma}\left\{\int_{\theta}e^{\imath\sigma\cdot[s+\imath\varphi(z,s)-s'-\imath\varphi(z,s')]-\varepsilon|\sigma|^2}f(z,s')\wedge dw\right\}d\sigma ;$$

(II.2.21) $\qquad\qquad\qquad K_{\mathcal{O},\Gamma}^{\varepsilon}f(z,s) =$

$$(2\pi)^{-m}\imath^n\int_{\Gamma}\left\{\int_{\mathcal{O}}e^{\imath\sigma\cdot(w-w')+a\cdot(z-z')-\varepsilon|\sigma|^2}f(z',s')\wedge E_a\wedge dw\right\}d\sigma ;$$

(II.2.22) $\qquad\qquad\qquad R_{\mathcal{O},\Gamma}^{\varepsilon}f(z,s) =$

$$(2\pi)^{-m}\imath^n\int_{\Gamma}\left\{\int_{\mathcal{O}}e^{\imath\sigma\cdot(w-w')+a\cdot(z-z')-\varepsilon|\sigma|^2}f(z',s')\wedge D_a\wedge dz\wedge dw\right\}d\sigma .$$

We also introduce the analogue of (II.2.21) with $\partial\mathcal{O}$ substituted for $\mathcal{O}$. We reach the following conclusion:

THEOREM II.2.3.– *Suppose* $0 \leq q \leq n$. *For any* $\varepsilon > 0$ *and any form* $f \in C^{\infty}(\mathcal{C}\!\angle\mathcal{O};$ $\Lambda^{m,q})$ *we have, in the open set* $\mathcal{O}$,

(II.2.23)

$$I^{\varepsilon}_{\theta,\Gamma}f \; = \; d[(-1)^q K^{\varepsilon}_{\mathcal{O},\Gamma}f] \; + \; (-1)^{d+q+1} K^{\varepsilon}_{\mathcal{O},\Gamma}df \; + \; (-1)^{d+q} K^{\varepsilon}_{\partial\mathcal{O},\Gamma}f \; + \; R^{\varepsilon}_{\mathcal{O},\Gamma}f \; .$$

We derive from Th. II.2.2:

THEOREM II.2.4.– *Let* q *be an integer,* $0 \leq q \leq n$. *Suppose* (II.2.8) *holds in* $\mathcal{O}\!\times\!\mathcal{O}\!\times\!\Gamma$. *Then, whatever the number* $\varepsilon > 0$ *and the differential form* $f \in C^{\infty}(\mathcal{C}\!\angle\mathcal{O};\Lambda^{m,q})$, *we have, in* $\mathcal{O}$,

(II.2.24) $\quad I^{\varepsilon}_{\theta,\Gamma}f \; = \; d[(-1)^q K^{\varepsilon}_{\mathcal{O},\Gamma}f] \; + \; (-1)^{d+q+1} K^{\varepsilon}_{\mathcal{O},\Gamma}df \; + \; (-1)^{d+q} K^{\varepsilon}_{\partial\mathcal{O},\Gamma}f \; .$

COROLLARY II.2.3.– *Suppose* (II.2.13) *holds in* $\mathcal{O}\!\times\!\mathcal{O}\!\times\!\Gamma$. *Then, for all numbers* $\varepsilon > 0$ *and all differential forms* $f \in C^{\infty}(\mathcal{C}\!\angle\mathcal{O};\Lambda^{m,0})$, *we have, in* $\mathcal{O}$,

(II.2.26) $\qquad\qquad (-1)^d I^{\varepsilon}_{\theta,\Gamma}f \; = \; - K^{\varepsilon}_{\mathcal{O},\Gamma}df \; + \; K^{\varepsilon}_{\partial\mathcal{O},\Gamma}f \; .$

COROLLARY II.2.4.– *Suppose* (II.2.14) *holds in* $\mathcal{O}\!\times\!\mathcal{O}\!\times\!\Gamma$. *Then, for all numbers* $\varepsilon > 0$ *and all differential forms* $f \in C^{\infty}(\mathcal{C}\!\angle\mathcal{O};\Lambda^{m,n})$, *we have, in* $\mathcal{O}$,

(II.2.27) $\qquad\qquad I^{\varepsilon}_{\theta,\Gamma}f \; = \; (-1)^m d K^{\varepsilon}_{\mathcal{O},\Gamma}f \; .$

II.3.   HOMOTOPY FORMULAS FOR FORMS THAT VANISH ON

        THE s-PART OF THE BOUNDARY

We return to Th. II.2.2. Recalling that $O = \Omega \times \theta$ we denote by $C^\infty_\bullet(\mathcal{E}\angle O; \Lambda^{m,q})$
the (closed) subspace of $C^\infty(\mathcal{E}\angle O; \Lambda^{m,q})$ consisting of those *forms f whose pull-back*
to to the "s-part" of the boundary, $\Omega \times \partial\theta$, *vanishes identically.* If $f \in C^\infty_\bullet(\mathcal{E}\angle O;$
$\Lambda^{m,q})$ then $K^a_{\partial O} f = K^a_{\partial\Omega \times \theta} f$ with the appropriate orientation of $\partial\Omega \times \theta$.

We shall avail ourselves of Th. I.3.2. We let now the map a depend not only
on (z,s,z',s') but also on $t \in [0,1]$. We are going to use the notation similar to that
in Sect. I.3 : $a_0 = a \big|_{t=0}$, $a_1 = a \big|_{t=1}$. And we introduce the notation (*cf.*
(I.3.16), (I.3.17), (I.3.18), (II.2.4), (II.2.5))

(II.3.1)

$$B^a_{\partial\Omega \times \theta} f (z,s) = (-2\iota\pi)^{-n} \int_{\partial\Omega \times \theta \times [0,1]} e^{a \cdot (z-z')} f(z',s') \wedge F_a \wedge dw \; ,$$

(II.3.2)

$$R^a_{\partial\Omega \times \theta} f (z,s) =$$

$$(-2\iota\pi)^{-n} \int_{\partial\Omega \times \theta \times [0,1]} e^{a \cdot (z-z')} f(z',s') \wedge D a \wedge dz \wedge dw \; ,$$

(II.3.3)

$$S^a_{\partial\Omega \times \theta} f (z,s) =$$

$$(-2\iota\pi)^{-n} \int_{\partial\Omega \times \theta} e^{a \cdot (z-z')} f(z',s') \wedge S_a \wedge dw \big|_{t=1} \; .$$

In what follows we shall agree that

(II.3.4) $K^{a_0}_{\partial\Omega \times \theta} f (z,s) = (-2\iota\pi)^{-n} \int_{\partial\Omega \times \theta} e^{a \cdot (z-z')} f(z',s') \wedge E_a \wedge dw \big|_{t=0} \; .$

By essentially duplicating the proof of Th. I.3.2 we derive, for any $1 \le q \le n-1$ and
any $f \in C^\infty_\bullet(\mathcal{E}\angle O; \Lambda^{m,q})$, in the open set O ,

(II.3.5) $- (-1)^{d+q} \mathcal{K}^{a}_{\partial\Omega\times\theta} f = d[(-1)^q B^{a}_{\partial\Omega\times\theta} f] + (-1)^{d+q+1} B^{a}_{\partial\Omega\times\theta} df +$

$$\mathcal{Z}^{a}_{\partial\Omega\times\theta} f - (-1)^{d+q} S^{a}_{\partial\Omega\times\theta} f .$$

We seek conditions ensuring that the operators $\mathcal{Z}^{a}_{\partial\Omega\times\theta}$ and $S^{a}_{\partial\Omega\times\theta}$ vanish identically. By analogy with the argument leading to Th. I.4.4 we see that $\mathcal{Z}^{a}_{\partial\Omega\times\theta} f \equiv 0$ provided, in the "cylinder"

$$\mathcal{O}\times\partial\Omega\times\theta\times[0,1] = \{ (z,s,z',s',t') ; z \in \Omega , z' \in \partial\Omega , s , s' \in \theta , 0 \leq t \leq 1 \} ,$$

we have, using the notation (II.2.9),

(II.3.6) $d_t a_i \wedge \left\{ \bigwedge_{\substack{j \in J \\ j \neq i}} L a_j \right\} \wedge \left\{ \bigwedge_{k \notin J} L' a_k \right\} \equiv 0 , \quad \forall J , |J| = q+1 , i \in J ;$

$S^{a}_{\partial\Omega\times\theta} f \equiv 0$ provided (II.2.10) holds in the "face" $t = 1$ of that same cylinder. Under these hypotheses we derive from (II.3.5), for any $f \in C^{\infty}_{\bullet}(\mathscr{E}\diagup\mathcal{O};\Lambda^{m,q})$,

$$- (-1)^{d+q} \mathcal{K}^{a}_{\partial\Omega\times\theta} f = d[(-1)^q B^{a}_{\partial\Omega\times\theta} f] + (-1)^{d+q+1} B^{a}_{\partial\Omega\times\theta} df .$$

We obtain (*cf.* Th. I.4.4):

THEOREM II.3.1.– *Suppose* $1 \leq q \leq n-1$ *and that the following hypotheses are satisfied:* (II.2.8) *in* $\mathcal{O}\times\mathcal{O}$ *when* $t = 0$; (II.2.10) *in* $\mathcal{O}\times\partial\Omega\times\theta$ *when* $t = 1$; (II.3.6) *in* $\mathcal{O}\times\partial\Omega\times\theta\times$ [0,1]. *Then, for any differential form* $f \in C^{\infty}_{\bullet}(\mathscr{E}\diagup\mathcal{O};\Lambda^{m,q})$,

(II.3.7) $\int_{\theta} f (z,s') \wedge dw =$

$$d[(-1)^{q}(\mathcal{K}^{a}_{\mathcal{O}} - B^{a}_{\partial\Omega\times\theta}) f] + (-1)^{d+q+1}(\mathcal{K}^{a}_{\mathcal{O}} - B^{a}_{\partial\Omega\times\theta}) df .$$

We shall also have use for the result analogous to Th. I.4.3. We introduce the condition in $\mathcal{O}\times\mathcal{O}\times[0,1]$,

$$(\mathrm{II}.3.8) \qquad d_t a_i \wedge \left\{ \bigwedge_{\substack{j\in J \\ j\neq i}} L\, a_j \right\} \wedge \left\{ \bigwedge_{k\notin J} L'\, a_k \right\} \equiv 0 \ , \quad \forall\, J \ , \ |J| = q \ , \ i \in J \ .$$

THEOREM II.3.2.– *Suppose* $1 \leq q \leq n-1$ *and that the following hypotheses are satisfied:* (II.2.10) *in* $\mathcal{O}\times\mathcal{O}$ *when* $t = 1$; (II.3.8) *in* $\mathcal{O}\times\mathcal{O}\times[0,1]$. *Then, for any closed differential form* $f \in C_\bullet^\infty(\mathscr{C}\angle\mathcal{O};\Lambda^{m,q})$,

$$(\mathrm{II}.3.9) \qquad \int_\theta f\,(z,s')\wedge dw \;=\; d[(-1)^q (K_{\mathcal{O}}^{\mathcal{A}} - B_{\partial\Omega\times\theta}^{a})f\,] \ .$$

The next step is to let a depend on $\sigma \in \mathbb{R}^d$ and to introduce the analogues of (II.2.21) & (II.2.22):

$$(\mathrm{II}.3.10) \qquad K_{\partial\Omega\times\theta,\Gamma}^{\varepsilon} f\,(z,s) =$$

$$(2\pi)^{-m}i^n \int_\Gamma \left\{ \int_{\partial\Omega\times\theta} e^{i\sigma\cdot(w-w') + a\,\cdot(z-z') - \varepsilon|\sigma|^2} f\,(z',s')\wedge E_a \wedge dw \big|_{t=0} \right\}\,d\sigma \ ,$$

$$(\mathrm{II}.3.11) \qquad B_{\partial\Omega\times\theta,\Gamma}^{\varepsilon} f\,(z,s) =$$

$$(2\pi)^{-m}i^n \int_\Gamma \left\{ \int_{\partial\Omega\times\theta\times[0,1]} e^{i\sigma\cdot(w-w') + a\,\cdot(z-z') - \varepsilon|\sigma|^2} f\,(z',s')\wedge F_a \wedge dw \right\}d\sigma \ .$$

We shall also make use of $K_{\Omega\times\partial\theta,\Gamma}^{\varepsilon} f$, which is defined like $K_{\partial\Omega\times\theta,\Gamma}^{\varepsilon} f$ except that the integration with respect to (z',s') is carried out over $\Omega\times\partial\theta$. Of course, $K_{\partial\Omega\times\theta,\Gamma}^{\varepsilon} f$ with the appropriate orientation of the integration domain, is equal to $K_{\partial\mathcal{O},\Gamma}^{\varepsilon} f$ if $f \in C_\bullet^\infty(\mathscr{C}\angle\mathcal{O};\Lambda^{m,q})$; $K_{\partial\mathcal{O},\Gamma}^{\varepsilon} f$ refers to the form (II.2.21) where the integration with respect to (z',s') is carried out over $\partial\mathcal{O}$ and where $a = a\,(z,s,z',s',0)$.

Th. II.3.1 has the following straight–forward consequence (*cf* Th. II.2.3):

THEOREM II.3.3.– *Suppose* $1 \leq q \leq n-1$, *and that the following is true:*

    ı)   (II.2.8) *holds for all* $(z,s,z',s') \in \mathcal{O} \times \mathcal{O}$, $\sigma \in \Gamma$, *when* $t = 0$;

    ıı)   (II.2.10) *holds for all* $(z,s) \in \mathcal{O}$, $z' \in \partial\Omega$, $s' \in \Theta$, $\sigma \in \Gamma$, *when* $t = 1$;

    ııı) (II.3.6) *holds for all* $(z,s) \in \mathcal{O}$, $z' \in \partial\Omega$, $s' \in \Theta$, $0 \leq t \leq 1$, $\sigma \in \Gamma$.

Then, for all numbers $\varepsilon > 0$ *and all differential forms* $f \in C_\bullet^\infty(\mathcal{C} \swarrow \mathcal{O}; \Lambda^{m,q})$ *we have, in the open set* $\mathcal{O}$,

(II.3.13)

$$\mathbf{I}_{\theta,\Gamma}^\varepsilon f = d[(-1)^q (\mathbf{K}_{\mathcal{O},\Gamma}^\varepsilon - \mathbf{B}_{\partial\Omega\times\theta,\Gamma}^\varepsilon)f] + (-1)^{d+q+1}(\mathbf{K}_{\mathcal{O},\Gamma}^\varepsilon - \mathbf{B}_{\partial\Omega\times\theta,\Gamma}^\varepsilon)df .$$

In (II.3.13) $\mathbf{K}_{\mathcal{O},\Gamma}^\varepsilon f$ is the form (II.2.21) in which $a = a(z,s,z',s',0)$.

We shall also need the CR version of Th. I.4.5:

THEOREM II.3.4.– *Suppose* $1 \leq q \leq n-1$ *and that the following is true:*

    ı)   (II.2.10) *holds for all* $(z,s,z,s') \in \mathcal{O} \times \mathcal{O}$, $\sigma \in \Gamma$, *when* $t = 1$;

    ıı)   (II.3.8) *holds for all* $(z,s,z',s') \in \mathcal{O} \times \mathcal{O}$, $0 \leq t \leq 1$, $\sigma \in \Gamma$.

Then, for all numbers $\varepsilon > 0$ *and all closed differential forms* $f \in C_\bullet^\infty(\mathcal{C} \swarrow \mathcal{O}; \Lambda^{m,q})$ *we have, in the open set* $\mathcal{O}$,

(II.3.14)
$$\mathbf{I}_{\theta,\Gamma}^\varepsilon f = d[(-1)^q (\mathbf{K}_{\mathcal{O},\Gamma}^\varepsilon - \mathbf{B}_{\partial\Omega\times\theta,\Gamma}^\varepsilon)f] .$$

II.4.   THE PINCHING TRANSFORMATION

We introduce the compact set

$$\gamma = \{ \ w \in \mathbf{C}^d \ ; \ \exists \ s \in \mathscr{C}\!\!\diagup\!\theta \ , \ w = s + \imath\varphi(0,s) \ \} \ .$$

We regard γ as a C^∞ manifold with boundary, $\partial\gamma = \{ \ w \in \mathbf{C}^d \ ; \ \exists \ s \in \partial\theta \ , \ w = s + \imath\varphi(0,s) \ \}$. The map $w \rightarrow (0,w)$ from $\mathbf{C}^d$ into $\mathbf{C}^{n+d}$ allows us to identify γ to a subset of the generic submanifold Σ defined by the equations (II.1.35) and its "interior", $\overset{\circ}{\gamma} = \{ \ w \in \mathbf{C}^d \ ; \ \exists \ s \in \theta \ , \ w = s + \imath\varphi(0,s) \ \}$, to a totally real d–dimensional submanifold of $\mathbf{C}^d$ on which the origin lies. Throughout the remainder of the present chapter we shall reason under the following hypothesis:

(II.4.1)   *There is a holomorphic function Q in an open neighborhood of γ in $\mathbf{C}^d$ such that $Q(w) \neq 0$ for every $w \in \overset{\circ}{\gamma}$ and $Q \equiv 0$ on $\partial\gamma$.*

We call the attention of the reader to the following facts ([B–T, 1988], Propositions 2.1, 2.2):

PROPOSITION II.4.1.– *If $d = 1$ and if θ is an open interval $]-a,b[$ (a , $b > 0$) then pro– perty* (II.4.1) *holds.*

Proof: It suffices to take $Q(w) = [w+a-\imath\varphi(0,-a)][w-b-\imath\varphi(0,b)]$. $\blacksquare$

PROPOSITION II.4.2.– *Suppose the function $s \rightarrow \varphi(0,s)$ is real–analytic. If θ is any open ball centered at the origin whose radius is sufficiently small, then* (II.4.1) *holds.*

Proof: Thanks to the analyticity of $\varphi(0,\cdot)$ and to hypothesis (II.1.3) there is a unique holomorphic solution $G(w)$, in a suitably small open neighborhood of the origin in $\mathbf{C}^d$, of the equations

$$w = G + i\varphi(0,G) , \ G(0) = 0 .$$

Note that $G(s+i\varphi(0,s)) = s$. Let $\theta = \{ \ s \in \mathbb{R}^d \ ; \ |s| < \delta \ \}$. If $\delta > 0$ is suffi-
ciently small the function

$$Q(w) = \delta^2 - G_1^2(w) - ... - G_d^2(w)$$

will be defined and holomorphic in an open neighborhood of γ . When $s \in \mathscr{C}\angle\theta$,
$Q(s+i\varphi(0,s)) = \delta^2 - |s|^2$. ∎

REMARK II.4.1.– It is easy to produce examples of smooth, but not analytic, maps φ
such that the conclusion in Prop. II.4.2 holds, for instance by suitably perturbing an
analytic map φ . ∎

REMARK II.4.2.– In [J] the reader will find examples of C^∞ maps φ which do not pos-
sess property (II.4.1) (regardless of the choice of θ). ∎

We shall make use of the function Q in (II.4.1) to define the map

(II.4.2) $(z,s) \to (z/Q(s+i\varphi(z,s)),s)$,

from an open subset $\hat{O}$ of $\mathbb{C}^n \times \mathbb{R}^d$ onto $O = \Omega \times \theta$. To start with we assume that Ω is
small enough that Q is holomorphic in an open neighborhood of the image of $\mathscr{C}\angle O$
under the map $(z,s) \to Q(s+i\varphi(z,s))$. This is permitted since $Q(w)$ is holomorphic in
an open neighborhood of the compact set γ . Next we note that, to each $s \in \theta$,
there is an open neighborhood of the origin in z–space $\mathbb{C}^n$, $\mathscr{D}(s) \subset \mathscr{D}$, such that

(II.4.3) $z \in \mathscr{D}(s) \Rightarrow Q(s+i\varphi(z,s)) \neq 0$.

We shall therefore require

(II.4.4) $$\hat{\mathcal{O}} \subset \{ (z,s) \in \mathbb{C}^n \times \Theta \ ; \ z \in \mathscr{D}(s) \} \ .$$

Further delimitation of $\hat{\mathcal{O}}$ is achieved thanks to the following

LEMMA II.4.1.– *If the open set* Ω *is sufficiently small there is a unique* C^∞ *solution* z $= H(\zeta,s)$ *in an open neighborhood of* $\mathscr{C} \angle \mathcal{O}$, $\mathcal{O}_1$, *of the equation*

(II.4.5) $$z - Q(s+\imath\varphi(z,s))\zeta = 0 \ ,$$

such that $H(0,s) \equiv 0$. *Furthermore, there are* C^∞ *functions* M , $N : \mathcal{O}_1 \to \mathbb{C}$, *of the form* $1 + O(|\zeta|)$ *and such that*

(II.4.6) $$H(\zeta,s) = [Q(s+\imath\varphi(0,s))M(\zeta,s) + \overline{Q}(s-\imath\varphi(0,s))N(\zeta,s)]\zeta \ .$$

Proof: The Jacobian determinant with respect to (x,y) of the left–hand side of (II. 4.5) is equal to $1 + O(|\zeta|)$. Thus the implicit function theorem applies, uniformly with respect to s in a compact neighborhood of $\mathscr{C} \angle \Theta$ in $\mathscr{B}$, as soon as diam Ω is sufficiently small. This gets us the solution $H(\zeta,s)$. The open neighborhood $\mathcal{O}_1 \subset \mathscr{D}$ $\times \mathscr{B}$ of $\mathscr{C} \angle \mathcal{O}$ can be chosen in such a way that

(II.4.7) $$Q(s+\imath\varphi(z,s)) = Q(s+\imath\varphi(0,s)) + z \cdot A(z,s) + \bar{z} \cdot B(z,s) \ ,$$

with $A = (A_1 ,...,A_n)$, $B = (B_1 ,...,B_n) \in C^\infty(\mathcal{O}_1;\mathbb{C}^n)$. Multiplying both sides of (II.4. 7) by $\zeta_j \ (j = 1,...,n)$ and replacing z by $H(\zeta,s)$ yields

$$H - [H \cdot A(H,s) + \overline{H} \cdot B(H,s)]\zeta = Q(s+\imath\varphi(0,s))\zeta \ .$$

Set $\chi_j = H_j / \zeta_j \ (j = 1,...,n)$, $A\circ\zeta = (A_1(H,s)\zeta_1,...,A_n(H,s)\zeta_n)$, $B\circ\bar{\zeta} = (B_1(H,s)\bar{\zeta}_1,...$ ..$, B_n(H,s)\bar{\zeta}_n)$; we get

$$\chi_j - (A \circ \zeta) \cdot \chi - (B \circ \overline{\zeta}) \cdot \overline{\chi} = \mathcal{Q}(s + \imath \varphi(0,s)) \ ,$$

(II.4.8)

$$\overline{\chi}_j - (B \circ \zeta) \cdot \chi - (\overline{A} \circ \overline{\zeta}) \cdot \overline{\chi} = \mathcal{Q}(s - \imath \varphi(0,s)) \ .$$

We may solve (II.4.8) as a system of $2n$ linear equations in the unknowns χ_i, $\overline{\chi}_j$. Notice that they imply $\chi_i = \chi_1$ for all $i = 2,...,n$. We get at once (II.4.6). ∎

COROLLARY II.4.1.– *If $s \in \partial\Theta$ then $H(\zeta,s) = 0$ for all $\zeta \in \Omega$.*

Proof: When $s \in \partial\Theta$, $\mathcal{Q}(s + \imath \varphi(0,s)) = 0$, whence the assertion, by (II.4.6). ∎

In the sequel we assume that $\mathrm{diam}\ \Omega$ is small enough to insure the validity of the conclusions in Lemma II.4.1. They have the consequence that the mapping

(II.4.9) $(\zeta,s) \rightarrow (H(\zeta,s),s)$

from $\mathcal{O}$ to $\hat{\mathcal{O}}$ (defined as the image of $\mathcal{O}$) is the inverse of the mapping (II.4.2). If $\mathrm{diam}\ \Omega$ is sufficiently small condition (II.4.4) will be automatically satisfied; (II.4. 9) will be a diffeomorphism of $\mathcal{O}$ onto $\hat{\mathcal{O}}$ (the image of $\mathcal{O}$) which extends as a C^∞ map from $\mathscr{C}\!\ell\,\mathcal{O}$ to $\mathscr{C}\!\ell\,\hat{\mathcal{O}}$. Note, however, that the extension is *not* injective: it maps the part $\Omega \times \partial\Theta$ of the boundary of $\mathcal{O}$ onto $\{0\} \times \partial\Theta$.

In the sequel we write

(II.4.10) $\tau = s + \imath \varphi(H(\zeta,s),s)$;

in other words, τ is the pull–back of w under the map (II.4.9). Note also that, according to (II.4.5),

(II.4.11) $H(\zeta,s) = \mathcal{Q}(\tau)\zeta$.

Let then $\hat{\Sigma}$ denote the generic submanifold of $\mathbb{C}^{n+d}$ defined by the conditions

(II.4.12) $(z, \mathscr{R}e\, w) \in \hat{O}$, $\mathscr{I}m\, w = \varphi(z, \mathscr{R}e\, w)$,

and let $\Sigma^{\#}$ denote the generic submanifold of $\mathbb{C}^{n+d}$ (where now the complex coordinates are ζ_j and w_k) defined by

(II.4.13) $(\zeta, \mathscr{R}e\, w) \in O$, $\mathscr{I}m\, w = \varphi(H(\zeta, \mathscr{R}e\, w), \mathscr{R}e\, w)$.

Observe that $\hat{\Sigma}$ is an open neighborhood of the origin in Σ and that the map

(II.4.14) $(z, w) \rightarrow (z/\mathcal{Q}(w), w)$

induces a diffeomorphism of $\hat{\Sigma}$ onto $\Sigma^{\#}$. The inverse diffeomorphism,

(II.4.15) $(\zeta, \tau) \rightarrow (\mathcal{Q}(\tau)\zeta, \tau)$,

extends as a C^{∞} map from the closure of $\Sigma^{\#}$ (viewed as a compact manifold with boundary) to the closure of $\hat{\Sigma}$ in Σ . It is evident that (II.4.15) extends as a biholomorphism of an open neighborhood of $\Sigma^{\#}$ in $\mathbb{C}^{n+d}$ onto one of $\hat{\Sigma}$.

Next we consider differential forms f in $\hat{O}$ given by (II.1.26); we assume that $f \in C^{\infty}(\mathscr{C}\!\ell\,\hat{O}; \Lambda^{m,q})$, i. e., $f_J \in C^{\infty}(\mathscr{C}\!\ell\,\hat{O})$, $\forall\, J$, $|J| = q$. We shall pull–back the form f to O by means of the map (II.4.9):

(II.4.16) $\wp^* f = dH(\zeta, s) \wedge d\tau \wedge \displaystyle\sum_{|J|=q} f_J(H(\zeta, s), s) d\overline{H(\zeta, s)}_J$.

It ought to be stressed that $\wp^* f \in C^{\infty}(\mathscr{C}\!\ell\,O; \Lambda^{m,q})$ due to the fact that the map (II.4. 9) extends as a C^{∞} map from $\mathscr{C}\!\ell\,O$ to $\mathscr{C}\!\ell\,\hat{O}$ and that, on O , it is equal to the restriction of the biholomorphism (II.4.15). Again, by (II.4.11), we have

$$(\text{II.4.17}) \qquad \wp^* f \;=\; \mathcal{Q}(\tau)^n \mathrm{d}\zeta \wedge \mathrm{d}\tau \wedge \sum_{|J|=q} f_J(\mathcal{Q}(\tau)\zeta, s)\,\mathrm{d}[\overline{\mathcal{Q}(\tau)\zeta}]_J \;.$$

PROPOSITION II.4.2.– *Let* $f \in C^\infty(\mathscr{E} \angle \hat{\mathcal{O}}; \Lambda^{m,q})$. *The pull–back of* $\wp^* f$ *to* $\Omega \times \partial\Theta$ *vanishes identically.*

Proof: By Cor. II.4.1 and (II.4.11) we have $\mathcal{Q}(\zeta, s)\zeta \equiv 0$ if $s \in \partial\Theta$ and $\zeta \in \Omega$. This is only possible if $\mathcal{Q}(\zeta, s) \equiv 0$ in $\Omega \times \partial\Theta$, whence the assertion, by (II.4.17). ∎

Finally consider the *push–forward* $\wp_* F$ of a form $F \in C^\infty(\mathscr{E} \angle \mathcal{O}; \Lambda^{m,q})$ under the map (II.4.16) (*i. e.*, the pull–back of F under the map (II.4.15)). If

$$(\text{II.4.18}) \qquad F \;=\; \mathrm{d}\zeta \wedge \mathrm{d}\tau \wedge \sum_{|J|=q} F_J(\zeta, s)\,\mathrm{d}\overline{\zeta}_J \;,$$

then

$$(\text{II.4.19}) \qquad \wp_* F \;=\; \mathcal{Q}(w)^{-n} \mathrm{d}z \wedge \mathrm{d}w \wedge \sum_{|J|=q} F_J(z/\mathcal{Q}(w), s)\,\mathrm{d}[\overline{z/\mathcal{Q}(w)}]_J \;.$$

In general the coefficients of $\wp_* F$ will blow up as s approaches the boundary $\partial\Theta$ (and $z \to 0$).

II.5   **REDUCTION TO FORMS THAT VANISH ON THE s-PART**

        OF THE BOUNDARY

As announced in Sect. II.4 we continue to reason under hypothesis (II.4.1).
From now on we also reason under the hypothesis that there is a number κ , $0 <$
$\kappa < 1$, such that, if we set

(II.5.1) $$\mathscr{C}_\kappa = \{\ \sigma \in \mathbb{C}^d \ ; \ \mathscr{R}e\ \sigma \in \Gamma \ , \ |\mathscr{I}m\ \sigma| < \kappa|\mathscr{R}e\ \sigma| \ \} \ ,$$

then

(II.5.2)   *the map* $\mathbf{\mathit{a}} : (\mathscr{D} \times \mathscr{B}) \times (\mathscr{D} \times \mathscr{B}) \times \Gamma \times [0,1] \to \mathbb{C}^n$ *extends as a* C^∞ *map*

                $(\mathscr{D} \times \mathscr{B}) \times (\mathscr{D} \times \mathscr{B}) \times \mathscr{C}_\kappa \times [0,1] \to \mathbb{C}^n,$ *holomorphic with respect to* σ ;

(II.5.3)   *to each set of multi–indices* $(\alpha, \overline{\alpha}, \beta, \alpha', \overline{\alpha}', \beta') \in \mathbb{Z}_+^n \times \mathbb{Z}_+^n \times \mathbb{Z}_+^d \times \mathbb{Z}_+^n \times$

                $\mathbb{Z}_+^n \times \mathbb{Z}_+^d$ *there is a constant* $C > 0$ *such that*

                $$| \partial_z^\alpha \partial_{\overline{z}}^{\overline{\alpha}} \partial_s^\beta \partial_{z'}^{\alpha'} \partial_{\overline{z}'}^{\overline{\alpha}'} \partial_{s'}^{\beta'} \mathbf{\mathit{a}} | \ \leq\ C|\sigma|$$

                in $(\mathscr{D} \times \mathscr{B}) \times (\mathscr{D} \times \mathscr{B}) \times \mathscr{C}_\kappa \times [0,1]$.

Notice that $\sigma = 0 \ \Rightarrow \ \mathbf{\mathit{a}} = 0$.

Otherwise we continue to use the notation of Sect. II.4. We are going to
apply Theorems II.2.4 & II.3.3 in (ζ,s)–space, to the CR structure $\mathcal{T}$ on $\mathcal{O}_1$ ($\mathcal{O}_1$ is an
open neighborhood of $\mathscr{C}\angle\mathcal{O}$; see Lemma II.4.1) defined by the functions ζ_j ($1 \leq j \leq$
n) and

(II.5.4) $$\tau_k = s_k + \imath \psi_k(\zeta,s) \ ,$$

where

$$\psi_k(\zeta,s) = \varphi_k(H(\zeta,s),s)$$

$[1 \leq k \leq d$; we write $\tau = (\tau_1,...,\tau_d)$, $\psi = (\psi_1,...,\psi_d)]$. It is important, however, that we continue to state the hypotheses in Theorems II.2.4 & II.3.3 in terms of the originally given CR structure. We must therefore show that these hypotheses, in the original structure, entail similar properties in the CR structure $\mathcal{T}$.

Let us introduce the pull-back of the map a :

$$\overset{\#}{a}(\zeta,s,\zeta',s',\sigma,t) = a\left(H(\zeta,s),s,H(\zeta',s'),s',\sigma,t\right) .$$

It is clear that

(II.5.5) $\overset{\#}{a} : \mathcal{O}_1 \times \mathcal{O}_1 \times \mathcal{C}_\kappa \times [0,1] \to \mathbb{C}^n$ is a C^∞ map, holomorphic with respect to σ , satisfying (II.5.3).

Below we use the holomorphic function $\mathcal{Q}_1$ defined by

(II.5.6) $\mathcal{Q}(\tau) - \mathcal{Q}(\tau') = \mathcal{Q}_1(\tau,\tau')\cdot(\tau-\tau')$.

The implicit function theorem implies that the equation

(II.5.7) $\sigma = G - \imath[\overset{\#}{a}(\zeta,s,\zeta',s',G,t)\cdot\zeta]\mathcal{Q}_1(\tau,\tau')$,

has a unique C^∞ solution G in $\mathcal{O}_1 \times \mathcal{O}_1 \times \mathcal{C}_{\kappa'} \times [0,1]$ such that $G\mid_{\sigma=0} = 0$, holomorphic with respect to σ , for a suitable choice of κ', $0 < \kappa' < \kappa < 1$. We avail ourselves of (II.5.3) by choosing $\mathcal{O}_1$ (and therefore $\mathcal{O}$) small enough that the image of $\mathcal{O}_1 \times \mathcal{O}_1 \times \mathcal{C}_{\kappa'} \times [0,1]$ under the map G is contained in $\mathcal{C}_\kappa$. This allows us to define the map $\tilde{a}$ which will play in (ζ,s)–space the role played by a in (z,s)–space:

(II.5.8) $\qquad \tilde{a}(\zeta,s,\zeta',s',\sigma,t) = Q(\tau') \overset{\#}{a}(\zeta,s,\zeta',s',G(\zeta,s,\zeta',s',\sigma,t),t)$.

The reason for such a choice of the map $\tilde{a}$ will become clear in the next section.

We begin by translating (II.2.8), or rather each one of the two properties (II. 2.10) or (II.2.11), assumed to be valid for all $(z,s,z',s',\sigma) \in \hat{O} \times \hat{O} \times \Gamma$ (and any value of t one deems necessary). Call $\mathcal{L}$ the differential operator analogous to L in the CR structure $\mathcal{T}$ (see (II.2.9)). By pull–back via the map $(\zeta,s) \to (H(\zeta,s),s)$ each vector field $\mathcal{L}_j$ is transformed into a linear combination of L_1 ,..., L_n . Due to this fact (II.2.10) entails

(II.5.9) $\qquad \bigwedge_{j \in J} \mathcal{L} \overset{\#}{a}_j \equiv 0$, $\forall J$, $|J| = q$,

for all $(\zeta,s,\zeta',s',\sigma,t) \in O \times O \times \Gamma \times [0,1]$. Note that, thanks to holomorphy, the validity of the last equation for all $\sigma \in \Gamma$ implies its validity for all $\sigma \in \mathscr{C}_\kappa$. This observation allows us to replace σ by $G(\zeta,s,\zeta',s',\sigma,t)$ in (II.5.9). On the other hand, by the chain rule applied to (II.5.8),

$$\mathcal{L} \tilde{a}(\zeta,s,\zeta',s',\sigma,t) = Q(\tau') \Big[(\mathcal{L} \overset{\#}{a})(\zeta,s,\zeta',s',G,t) +$$

$$\partial_\sigma \overset{\#}{a}(\zeta,s,\zeta',s',G,t) \cdot \mathcal{L}G \Big] .$$

Letting $\mathcal{L}$ act on both sides of (II.5.7) yields

$$\mathcal{L}G - \imath \{ [\partial_\sigma \overset{\#}{a}(\zeta,s,\zeta',s',G,t) \cdot \zeta] \cdot \mathcal{L}G \} Q_1(\tau,\tau') =$$

$$\imath [(\mathcal{L} \overset{\#}{a})(\zeta,s,\zeta',s',G,t) \cdot \zeta] Q_1(\tau,\tau') .$$

This entails that the one–forms $\mathcal{L}G_k$ ($k = 1,...,d$) and therefore also the forms

$$\partial_\sigma \tilde{d}_i^\#(\zeta,s,\zeta',s',G,t) \cdot \mathcal{L}G \ ,$$

as well as the one-forms $\mathcal{L}\tilde{a}_i$ ($1 \le i \le n$), are linear combinations, with smooth coefficients, of the one-forms

$$(\mathcal{L}\tilde{d}_j^\#)(\zeta,s,\zeta',s',G,t) \ , \ j = 1,...,n \ .$$

Combining this with (II.5.9) yields

$$\bigwedge_{j \in J} \mathcal{L}\tilde{a}_j \equiv 0 \ , \ \forall \ J \ , \ |J| = q \ ,$$

which is the sought translation of condition (II.2.10). Substituting $\mathcal{L}'$ for $\mathcal{L}$ and keeping in mind that $\mathcal{L}'\tau' = 0$ enables us to likewise translate (II.2.11). As a consequence, we have proved that the validity of the equation

(II.5.10) $$\left\{ \bigwedge_{j \in J} L \, a_j \right\} \wedge \left\{ \bigwedge_{k \notin K} L' \, a_k \right\} \equiv 0 \ , \ \forall \ J \ , \ K \ , \ |J| = |K| = q \ ,$$

for all $(z,s,z',s',\sigma) \in \hat{O} \times \hat{O} \times \Gamma$, entails

$$\left\{ \bigwedge_{j \in J} \mathcal{L}\tilde{a}_j \right\} \wedge \left\{ \bigwedge_{k \notin K} \mathcal{L}'\tilde{a}_k \right\} \equiv 0 \ , \ \forall \ J \ , \ K \ , \ |J| = |K| = q \ ,$$

for all $(\zeta,s,\zeta',s',\sigma) \in O \times O \times \Gamma$.

Next we turn our attention to condition (II.3.6), or rather a strengthend version of it, assumed to hold for all (z,s,z',s',σ,t) in $\hat{O} \times \partial\hat{O} \times \Gamma \times [0,1]$ with the proviso that $s' \notin \partial\Theta$:

(II.5.11) $$d_t a_i \wedge \left\{ \bigwedge_{j \in J} L \, a_j \right\} \wedge \left\{ \bigwedge_{k \notin K} L' \, a_k \right\} \equiv 0 \ ,$$

$$\forall \; i = 1,...,n \; , \; \forall \; J \; , \; K \; , \; |J| \; = \; q \; , \; |K| \; = \; q{+}1 \; .$$

Pulling back yields

$$d_t \mathcal{A}_i^{\#} \wedge \left\{ \bigwedge_{j \in J} \mathcal{L} \, \mathcal{A}_j^{\#} \right\} \wedge \left\{ \bigwedge_{k \notin K} \mathcal{L}' \, \mathcal{A}_k^{\#} \right\} \equiv 0 \; ,$$

for the same indices i and multi–indices J , K , now in the cylinder $\mathcal{O} \times \partial \Omega \times \Theta \times \Gamma \times$ $[0,1]$. We derive from (II.5.8):

$$d_t \tilde{a} \, (\zeta,s,\zeta',s',\sigma,t) \; = \; \mathcal{Q}(\tau') \Big[(d_t \mathcal{A}^{\#})(\zeta,s,\zeta',s',G \; ,t) \; +$$

$$\partial_\sigma \mathcal{A}^{\#}(\zeta,s,\zeta',s',G \; ,t) \cdot d_t G \Big] \; ,$$

and from (II.5.7):

$$d_t G \; - \; \imath \{ [\partial_\sigma \mathcal{A}^{\#}(\zeta,s,\zeta',s',G \; ,t) \cdot \zeta] \cdot d_t G \} \mathcal{Q}_1(\tau,\tau') \; =$$

$$\imath [d_t \mathcal{A}^{\#}(\zeta,s,\zeta',s',G \; ,t) \cdot \zeta] \mathcal{Q}_1(\tau,\tau') \; .$$

This shows that, for all $i = 1,...,n$, the one–forms

$$\partial_\sigma \mathcal{A}_i^{\#}(\zeta,s,\zeta',s',G \; ,t) \cdot d_t G \; ,$$

and hence also the forms $d_t \tilde{a}_i$, are linear combinations of the one–forms

$$d_t \mathcal{A}_j^{\#}(\zeta,s,\zeta',s',G \; ,t) \; , \; j = 1,...,n \; .$$

Here we reach the conclusion that, whatever the indices i and the multi–indices J K , as in (II.5.11),

$$d_t \tilde{a}_i \wedge \left\{ \bigwedge_{j \in J} \mathcal{L} \tilde{a}_j \right\} \wedge \left\{ \bigwedge_{k \notin K} \mathcal{L}' \tilde{a}_k \right\} \equiv 0$$

in the cylinder $\mathcal{O} \times \partial \Omega \times \theta \times \Gamma \times [0,1]$. A similar argument applies to the following strengthened version of (II.3.8):

(II.5.12)
$$d_t a_i \wedge \left\{ \bigwedge_{j \in J} L a_j \right\} \wedge \left\{ \bigwedge_{k \notin K} L' a_k \right\} \equiv 0 \ ,$$

$$\forall \ i = 1,...,n \ , \ \forall \ J \ , \ K \ , \ |J| = q{-}1 \ , \ |K| = q \ .$$

We are now in a position to apply Corollaries II.2.3, II.2.4 and also, thanks to Prop. II.4.2, Theorems II.3.3 & II.3.4. The operators $\Upsilon^\varepsilon_{\theta,\Gamma}$, $\check{K}^\varepsilon_{\mathcal{O},\Gamma}$ and $\check{B}^\varepsilon_{\partial \Omega \times \theta,\Gamma}$ are defined according to formulas (II.2.21) & (II.3.11), but with ψ substituted for φ and $\tilde{a}$ for a . We use the pull-back and push-forward maps $\wp^*$ & $\wp_*$ introduced in Sect. II.4; they commute with the exterior derivative, d :

THEOREM II.5.1.– *Suppose* (II.2.13) *holds in* $\hat{\mathcal{O}} \times \hat{\mathcal{O}} \times \Gamma$. *Then, for all numbers* $\varepsilon > 0$ *and all differential forms* $f \in C^\infty(\mathscr{C}\ell\hat{\mathcal{O}}; \Lambda^{m,0})$, *we have, in* $\hat{\mathcal{O}}$,

(II.5.13)
$$(-1)^d \wp_* \Upsilon^\varepsilon_{\theta,\Gamma} \wp^* f \ = \ -\wp_* \check{K}^\varepsilon_{\mathcal{O},\Gamma} \wp^* df \ + \ \wp_* \check{K}^\varepsilon_{\partial \Omega \times \theta,\Gamma} \wp^* f \ .$$

THEOREM II.5.2.– *Suppose* (II.2.14) *holds in* $\hat{\mathcal{O}} \times \hat{\mathcal{O}} \times \Gamma$. *Then, for all numbers* $\varepsilon > 0$ *and all differential forms* $f \in C^\infty(\mathscr{C}\ell\hat{\mathcal{O}}; \Lambda^{m,n})$, *we have, in* $\hat{\mathcal{O}}$,

(II.5.14)
$$\wp_* \Upsilon^\varepsilon_{\theta,\Gamma} \wp^* f \ = \ (-1)^m d[\wp_* \check{K}^\varepsilon_{\mathcal{O},\Gamma} \wp^* f] \ .$$

THEOREM II.5.3.– *Suppose* $1 \leq q \leq n{-}1$, *and that the following is true:*

  ı)   (II.5.10) *holds for all* $(z,s,z',s') \in \hat{\mathcal{O}} \times \hat{\mathcal{O}}$, $\sigma \in \Gamma$, *when* $t = 0$;

  ıı) (II.2.10) *holds for all* $(z,s,z',s') \in \hat{\mathcal{O}} \times \hat{\mathcal{O}}$, $\sigma \in \Gamma$, *when* $t = 1$;

  ııı) (II.5.11) *holds for all* $(z,s,z',s') \in \hat{\mathcal{O}} \times \partial \hat{\mathcal{O}}$, $s' \in \theta$, $0 \leq t \leq 1$, $\sigma \in \Gamma$.

Then, for all numbers $\varepsilon > 0$ *and all differential forms* $f \in C^\infty(\mathscr{C}\ell\,\hat{O};\Lambda^{m,q})$ *we have, in the open set* $\hat{O}$,

$$(\mathrm{II}.5.15) \qquad \wp_* \mathbf{I}^\varepsilon_{\theta,\Gamma} \wp^* f = \mathrm{d}[(-1)^q \wp_* (\mathbf{K}^\varepsilon_{\hat{O},\Gamma} - \mathbf{B}^\varepsilon_{\partial\Omega\times\theta,\Gamma}) \wp^* f] +$$

$$(-1)^{d+q+1} \wp_* (\mathbf{K}^\varepsilon_{\hat{O},\Gamma} - \mathbf{B}^\varepsilon_{\partial\Omega\times\theta,\Gamma}) \wp^* \mathrm{d} f .$$

THEOREM II.5.4.– *Suppose* $1 \le q \le n{-}1$ *and that the following is true:*

    $\imath)$ (II.2.10) *holds for all* $(z,s,z,s') \in \hat{O}\times\hat{O}$, $\sigma \in \Gamma$, *when* $t = 1$;

    $\imath\imath)$ (II.5.12) *holds for all* $(z,s,z',s') \in \hat{O}\times\hat{O}$, $0 \le t \le 1$, $\sigma \in \Gamma$.

Then, for all numbers $\varepsilon > 0$ *and all closed differential forms* $f \in C^\infty(\mathscr{C}\ell\,\hat{O};\Lambda^{m,q})$ *we have, in the open set* $\hat{O}$,

$$(\mathrm{II}.5.16) \qquad \wp_* \mathbf{I}^\varepsilon_{\theta,\Gamma} \wp^* f = \mathrm{d}[(-1)^q \wp_* (\mathbf{K}^\varepsilon_{\hat{O},\Gamma} - \mathbf{B}^\varepsilon_{\partial\Omega\times\theta,\Gamma}) \wp^* f] .$$

6.   CONVERGENCE OF THE HOMOTOPY OPERATORS

Finally we must address the question of the convergence, as $\varepsilon \to +0$, of the operators $\Upsilon^{\varepsilon}_{\theta,\Gamma}$, $\mathring{K}^{\varepsilon}_{O,\Gamma}$, $\mathring{K}^{\varepsilon}_{\partial\Omega\times\theta,\Gamma}$ and $\mathring{B}^{\varepsilon}_{\partial\Omega\times\theta,\Gamma}$. We shall limit our attention to the case

$$(\text{II.6.1}) \qquad \Gamma = \mathbb{R}^d\backslash\{0\} \ \ or \ else \ d = 1 \ and \ \Gamma = \mathbb{R}_+ \ \ or \ \Gamma = \mathbb{R}_- \ .$$

We continue to reason under hypothesis (II.4.1). Since the function Q may always be replaced by any one of its powers, in particular by Q^2, there is no loss of generality in also assuming that

$$(\text{II.6.2}) \qquad |Q(w)| \leq const.[\text{dist}(w,\partial\gamma)]^2 \ , \quad w \in \gamma$$

[recall that γ is the image of $\mathscr{C}\angle\theta$ under the map $s \to s+\imath\varphi(0,s)$]. Hypothesis (II.6.2) will be helpful in ensuring the convergence of $\mathring{K}^{\varepsilon}_{O,\Gamma}$, $\mathring{K}^{\varepsilon}_{\partial\Omega\times\theta,\Gamma}$ and $\mathring{B}^{\varepsilon}_{\partial\Omega\times\theta,\Gamma}$.

We begin by looking at $\Upsilon^{\varepsilon}_{\theta,\Gamma}$. We are going to make use of the notation

$$\mathcal{J}(\zeta,s,s') = I + \imath\int_0^1 \psi_s(\zeta,s'+\lambda(s-s'))\mathrm{d}\lambda$$

($I = d\times d$ identity matrix); our original hypothesis $\|\mathrm{d}\varphi\| << 1$ entails $\|\mathcal{J}-I\| << 1$.

LEMMA II.6.1.– *Assume* (II.6.1) *holds. Then, for any* $f \in C^{\infty}(\mathscr{C}\angle O;\Lambda^{m,q})$, *as* $\varepsilon \to +0$, $\Upsilon^{\varepsilon}_{\theta,\Gamma}f$ *converges in* $C^{\infty}(O;\Lambda^{m,q})$. *When* $\Gamma = \mathbb{R}^d\backslash\{0\}$ *its limit is equal to* f . *When* $d = 1$ *and* $\Gamma = \mathbb{R}_+$ (*resp.,* $\mathbb{R}_-$) *its limit is equal to*

$$\Upsilon^+_{\theta}f\,(\zeta,s) = -\int_0^{+\infty}\left\{\int_{\theta}e^{\imath\sigma(s-s')}\ \mathcal{J}(\zeta,s,s')^{-1}f\,(\zeta,s')\wedge\mathrm{d}\tau\right\}\ \mathrm{d}\sigma/2\pi$$

$$(resp., \quad \Upsilon_\theta^- f\,(\zeta,s) = -\int_{-\infty}^0 \left\{ \int_\theta e^{i\sigma(s-s')}\,\mathcal{J}(\zeta,s,s')^{-1} f\,(\zeta,s')\wedge d\tau \right\}\, d\sigma/2\pi \;.$$

Proof: We write (*cf.* (II.1.26), (II.2.1))

$$f\,(\zeta,s')\wedge d\tau \equiv (-1)^d d\zeta \wedge d\tau \wedge \sum_{|J|=q} d\bar{\zeta}_J \wedge f_J(\zeta,s')[\det(I + i\psi_s(\zeta,s'))]ds'$$

modulo terms in which at least one factor ds'_k is missing. It follows that $\Upsilon_{\theta,\Gamma}^\epsilon f\,(\zeta,s)$ is a differential form like f whose coefficients are given by

$$(2\pi)^{-d}\int_\Gamma \int_\theta e^{i\sigma\cdot[s+i\psi(\zeta,s)-s'-i\psi(\zeta,s')]-\epsilon\sigma\cdot\sigma}\, f_J(\zeta,s')[\det(I+i\psi_s(\zeta,s'))]ds'd\sigma =$$

$$(2\pi)^{-d}\int_\Gamma \int_\theta e^{i\sigma\cdot\mathcal{J}(\zeta,s,s')(s-s')-\epsilon\sigma\cdot\sigma}\, f_J(\zeta,s')[\det(I+i\psi_s(\zeta,s'))]ds'd\sigma \;.$$

Hypothesis (II.6.1) and the fact that $\|\mathcal{J} - I\| << 1$ allow us to make the change of σ–variable (or deformation of the contour of σ–integration) $\sigma \to {}^t\mathcal{J}(\zeta,s,s')^{-1}\sigma$. The convergence becomes apparent (by a standard argument of pseudodifferential operator theory, if one likes).

When $\Gamma = \mathbb{R}^d\backslash\{0\}$, the limit of the preceding integrals is equal to

$$(2\pi)^{-d}\int_{\mathbb{R}^d} \int_\theta e^{i\sigma\cdot(s-s')} f_J(\zeta,s')\det[\mathcal{J}(\zeta,s,s')^{-1}(I+i\psi_s(\zeta,s'))]ds'd\sigma =$$

$$f_J(\zeta,s)\; \det[\mathcal{J}(\zeta,s,s)^{-1}(I+i\psi_s(\zeta,s))] = f_J(\zeta,s) \;.\blacksquare$$

COROLLARY II.6.1.– *Suppose* $d = 1$. *For any* $f \in C^\infty(\mathscr{C}\ell\mathcal{O};\Lambda^{m,q})$ *re have, in* $\mathcal{O}$,

(II.6.3) $$f = \Upsilon_\theta^+ f + \Upsilon_\theta^- f\;.$$

The convergence of the operators $\mathcal{K}_{\mathcal{O},\Gamma}^\epsilon$, $\mathcal{K}_{\partial\Omega\times\theta,\Gamma}^\epsilon$ and $\mathcal{B}_{\partial\Omega\times\theta,\Gamma}^\epsilon$ requires a new kind of hypothesis, basic to the approach we are following. Here also, as in Sect. II.5, we want to state it in terms of the original structure in $\mathscr{D} \times \mathscr{B}$, defined by the functions z_i ($1 \leq i \leq n$) and $w_k = s_k + i\varphi_k(z,s)$ ($1 \leq k \leq d$). To do so we introduce a map

$$(II.6.4) \qquad \mu : (\mathscr{D} \times \mathscr{B}) \times (\mathscr{D} \times \mathscr{B}) \times \Gamma \times [0,1] \to GL(d,\mathbb{C})$$

[$GL(d,\mathbb{C})$: the group of nonsingular $d \times d$ complex matrices] submitted to the following requirements:

(II.6.5)   *the map μ is Lipschitz continuous and there are positive constants C,
        κ', $\kappa' < \text{Min}(1,\kappa)$, such that, in $(\mathscr{D} \times \mathscr{B}) \times (\mathscr{D} \times \mathscr{B}) \times \Gamma \times [0,1]$,*

$$|v - \mu v| \leq \kappa' |\mathscr{R}e\, \mu v| \;, \; \forall\, v \in \mathbb{R}^d \;.$$

Notice that the following are consequences of the inequality in (II.6.5):

$$|\mathscr{I}m\, \mu v| \leq \kappa' |\mathscr{R}e\, \mu v| \;, \; \forall\, v \in \mathbb{R}^d \;; \; \|\mu\| \leq (1-\kappa')^{-1} \;, \; \|\mu^{-1}\| \leq 1+\kappa' \;.$$

The central hypothesis of the present work, alluded to above, can now be stated:

(II.6.6)   *There is a map (II.6.4), endowed with property (II.6.5), and a constant
        $C_0 > 0$ such that*

$$\mathscr{I}m \left[\mu(z,s,z',s',\sigma,t)\sigma \cdot [s + i\varphi(z,s) - s' - i\varphi(z',s')] \; - \right.$$

$$\left. i\, a\, (z,s,z',s',\mu(z,s,z',s',\sigma,t)\sigma,t) \cdot (z-z') \right] \geq -\, C_0$$

$$\forall\; (z,s,z',s',\sigma) \in \mathring{O} \times \mathring{O} \times \Gamma \;.$$

In (II.6.6) t is any number, $0 \leq t \leq 1$; in applications below, we shall take $t = 0$.

Our first task is to show that (II.6.6) entails a similar property in (ζ,s)–space. Notice that the pull–back, under the map $(\zeta,s) \rightarrow (H(\zeta,s),s)$, enables to derive from (II.6.6):

(II.6.7)
$$\mathcal{I}m \left[\mu^{\#}(\zeta,s,\zeta',s',\sigma,t)\sigma \cdot [s + \imath\psi(\zeta,s) - s' - \imath\psi(\zeta',s')] \; - \right.$$

$$\left. \imath d^{\#}(\zeta,s,\zeta',s',\mu^{\#}(\zeta,s,\zeta',s',\sigma,t)\sigma,t)\cdot(H(\zeta,s) - H(\zeta',s')) \right] \; \geq \; - \; C_0 \; .$$

We avail ourselves of the fact that, by (II.4.11) & (II.5.6),

$$H(\zeta,s) \; - \; H(\zeta',s') \; = \; Q(\tau')(\zeta - \zeta') \; + \; [Q(\tau) - Q(\tau')]\zeta \; =$$

$$Q(\tau')(\zeta - \zeta') \; + \; [Q_1(\tau,\tau') \cdot (\tau - \tau')]\zeta \; .$$

Putting this into (II.6.7) yields

(II.6.8)
$$\mathcal{I}m \left[\{\mu^{\#}\sigma - \imath[\, d^{\#}(\zeta,s,\zeta',s',\mu^{\#}\sigma,t) \cdot \zeta]Q_1(\tau,\tau')\} \cdot (\tau - \tau') \; - \right.$$

$$\left. \imath Q(\tau') \, d^{\#}(\zeta,s,\zeta',s',\mu^{\#}\sigma,t) \cdot (\zeta - \zeta') \right] \; \geq \; - \; C_0 \; .$$

This leads us to define the map

$$\tilde{\mu} \; : \; \mathcal{O}_1 \times \mathcal{O}_1 \times \Gamma \times [0,1] \quad \rightarrow \quad GL(d,\mathbb{C})$$

by the following equation (cf. (II.5.7))

(II.6.9)
$$\tilde{\mu}\sigma \; = \; \mu^{\#}\sigma \; - \; \imath[\, d^{\#}(\zeta,s,\zeta',s',\mu^{\#}\sigma,t) \cdot \zeta]Q_1(\tau,\tau') \; .$$

The matrix $\tilde{\mu}$ is as close to μ as we wish, provided diam Ω is sufficiently small; and thus $\tilde{\mu}$ will also satisfy (II.6.5). The solution of (II.5.7) shows that

$$\mu^{\#}\sigma = G\left(\zeta,s,\zeta',s',\tilde{\mu}\sigma\right)$$

and therefore, with the definition (II.5.8), the inequality (II.6.8) can be rewritten as

(II.6.10) $\mathcal{I}m\left[\tilde{\mu}\sigma\cdot(\tau-\tau') - \imath\tilde{a}\left(\zeta,s,\zeta',s',\tilde{\mu}\sigma,t\right)\cdot(\zeta-\zeta')\right] \geq -C_0$,

for all $(\zeta,s,\zeta',s',\sigma) \in \mathcal{O}\times\mathcal{O}\times\Gamma$, which is what we wanted.

As we shall see, hypothesis (II.6.6), if valid when $t = 0$, ensures the convergence of $\overset{\varepsilon}{K}_{\mathcal{O},\Gamma}$ as $\varepsilon \to +0$. That of $\overset{\varepsilon}{B}_{\partial\Omega\times\theta,\Gamma}$ follows from a variant of (II.6.6):

(II.6.11)   *There is an open neighborhood of the origin in* $\mathbb{C}^n\times\mathbb{R}^d$, $\hat{\mathcal{O}}_0 \subset \hat{\mathcal{O}}$,
          and a constant $C_0 > 0$ *such that*

$$\mathcal{I}m\left[\mu(z,s,z',s',\sigma,t)\sigma\cdot[s+\imath\varphi(z,s)-s'-\imath\varphi(z',s')] - \right.$$

$$\left. \imath a\left(z,s,z',s',\mu(z,s,z',s',\sigma,t)\sigma,t\right)\cdot(z-z')\right] \geq -C_0 \ ,$$

$$\forall \ (z,s,z',s',\sigma,t) \in \hat{\mathcal{O}}_0\times\partial\hat{\mathcal{O}}\times\Gamma \ , \ s' \notin \partial\theta \ .$$

Exactly the same argument as the one applied to (II.6.6) enables to conclude that (II.6.10) is valid for all $(\zeta,s,\zeta',s',\sigma,t) \in \mathcal{O}_0\times\partial\Omega\times\theta\times\Gamma\times[0,1]$, where $\mathcal{O}_0$ is the image of $\hat{\mathcal{O}}_0$ under the map $(z,s) \to (z/\mathcal{Q}(w),s)$.

LEMMA II.6.2.– *Assume* (II.6.1) *is valid. Let the maps* a *&* μ *(see* (II.5.2), (II.6.4)) *have properties* (II.5.3) *&* (II.6.5) *respectively. Let* $\tilde{a}$ *be defined according to* (II.5.8). *If* diam Ω *is sufficiently small the following is true:*

If (II.6.6) *is valid at* $t = 0$ *and if* $\varepsilon \to +0$, *then, given any differential form* $f \in C^{\infty}(\mathcal{C}\ell\,\mathcal{O};\Lambda^{m,q})$, *the form* $\overset{\varepsilon}{K}_{\mathcal{O},\Gamma}f$ *(resp.,* $\overset{\varepsilon}{K}_{\partial\Omega\times\theta,\Gamma}f$ *) converges in* $C^{\infty}(\mathcal{O};\Lambda^{m,q-1})$

(*resp.*, $C^\infty(\mathcal{O};\Lambda^{m,q})$).

If (II.6.11) *is valid then the form* $\check{\mathrm{B}}^\varepsilon_{\partial\Omega\times\theta,\Gamma}f$ *converges in* $C^\infty(\mathcal{O}_0;\Lambda^{m,q-1})$.

$\mathscr{P}\!\mathit{roof}$: Let us prove the statement for

$$\check{\mathrm{K}}^\varepsilon_{\mathcal{O},\Gamma}f\,(\zeta,s) = (2\pi)^{-m}\imath^n\int_\Gamma\left\{\int_{\mathcal{O}}e^{\imath\sigma\cdot(\tau-\tau')+\tilde{a}\cdot(\zeta-\zeta')-\varepsilon|\sigma|^2}f\,(\zeta',s')\wedge E_{\tilde{a}}\wedge\mathrm{d}\tau\right\}\mathrm{d}\sigma\;;$$

the treatement of $\check{\mathrm{K}}^\varepsilon_{\partial\Omega\times\theta,\Gamma}$ & $\check{\mathrm{B}}^\varepsilon_{\partial\Omega\times\theta,\Gamma}f$ is similar (and actually slightly easier).

Let $\hat{\mathrm{M}}_k$ be the vector fields in (ζ,s)–space analogous to (II.1.6). We consider the second–order differential operator

$$\Delta_{\hat{\mathrm{M}}} = \hat{\mathrm{M}}_1^2 +...+ \hat{\mathrm{M}}_d^2\;.$$

We observe that

$$(1-\Delta_{\hat{\mathrm{M}}})e^{\imath\sigma\cdot(\tau-\tau')+\tilde{a}\cdot(\zeta-\zeta')} = (1+|\sigma|^2+\tilde{\mathscr{U}})e^{\imath\sigma\cdot(\tau-\tau')+\tilde{a}\cdot(\zeta-\zeta')}\;,$$

where

$$\tilde{\mathscr{U}} = -\Delta_{\hat{\mathrm{M}}}\tilde{a}\cdot(\zeta-\zeta') - 2\imath\sum_{k=1}^d\sigma_k\hat{\mathrm{M}}_k\tilde{a}\cdot(\zeta-\zeta') - \sum_{k=1}^d[\hat{\mathrm{M}}_k\tilde{a}\cdot(\zeta-\zeta')]^2\;.$$

As a consequence of (II.5.3) we see that

$$|\tilde{\mathscr{U}}| \leq const.|\sigma||\zeta-\zeta'|(1+|\sigma|)\;.$$

We take diam Ω so small that

$$|\tilde{\mathscr{U}}| \leq (1+ \mathscr{R}e<\sigma>^2)/2\;,\; \forall\;(\zeta,s,\zeta',s',\sigma) \in (\mathscr{C}\!\mathit{l\mathcal{O}})\times(\mathscr{C}\!\mathit{l\mathcal{O}})\times\mathscr{C}_{\kappa'}\;,$$

which **enables** us to introduce the second–order differential operator

$$u \to \mathcal{P}u = (1-\Delta_{\hat{M}})[(1+<\sigma>^2+\tilde{\varkappa})^{-1}u] \ .$$

We denote by $\mathcal{P}'$ the differential operator defined in the same manner, but by means of the vector fields $\hat{M}_k$ in (ζ',s')–space. Note that $\mathcal{P}'$ only involves partial derivatives with respect to the variables s' ; its coefficients are C^∞ functions of $(\zeta,s,\zeta',s',\sigma)$ in $(\mathcal{B}\angle 0){\times}(\mathcal{B}\angle 0){\times}\mathcal{B}_{\kappa'}$ holomorphic with respect to σ . We take advantage of the fact that each vector field $\hat{M}_k$ is equal to its own transpose with respect to the measure $d\tau$ (in s–space $\mathbb{R}^d$); integration by parts with respect to s' yields

(II.6.12)
$$\int_\Gamma \left\{ \int_0^\infty e^{\imath\sigma\cdot(\tau-\tau')+\tilde{a}\cdot(\zeta-\zeta')-\varepsilon|\sigma|^2} f(\zeta',s')\wedge E_{\tilde{a}} \right\} d\sigma =$$

$$\int_\Gamma \left\{ \int_0^\infty e^{\imath\sigma\cdot(\tau-\tau')+\tilde{a}\cdot(\zeta-\zeta')-\varepsilon|\sigma|^2} \mathcal{P}'^N [f(\zeta',s')\wedge E_{\tilde{a}}] \right\} d\sigma \ +$$

$$\int_\Gamma \left\{ \int_{\Omega\times\partial\Theta} e^{\imath\sigma\cdot(\tau-\tau')+\tilde{a}\cdot(\zeta-\zeta')-\varepsilon|\sigma|^2} \mathcal{R}'_N [f(\zeta',s')\wedge E_{\tilde{a}}] \right\} d\sigma \ ,$$

with the differential operators $\mathcal{P}'$, $\mathcal{R}'_N$ acting coefficientwise on the differential form $f(\zeta',s')\wedge E_{\tilde{a}}$ (in the basis consisting of the $d\zeta_i$, $d\bar{\zeta}_j$, $d\tau_k$ and of the $d\zeta'_i$, $d\bar{\zeta}'_j$, $d\tau'_k$). Provided $\kappa' > 0$ is small enough,

(II.6.13)   *to every $(\beta,\beta',N) \in \mathbb{Z}_+^d{\times}\mathbb{Z}_+^d{\times}\mathbb{Z}_+$ there is a constant $C_{\beta,\beta',N} > 0$ such that, in $(\mathcal{B}\angle 0){\times}(\mathcal{B}\angle 0){\times}\mathcal{B}_{\kappa'}$,*

$$\left| \partial_s^\beta \partial_{s'}^{\beta'} \left[\mathcal{P}'^N [f(\zeta',s')\wedge E_{\tilde{a}}] \right] \right| \leq C_{\beta,\beta',N} (1+|\sigma|^2)^{-N} |\zeta-\zeta'|^{-2n+1} \ ,$$

$$\left| \partial_s^\beta \partial_{s'}^{\beta'} \left[\mathcal{R}'_N [f(\zeta',s')\wedge E_{\tilde{a}}] \right] \right| \leq C_{\beta,\beta',N} |\zeta-\zeta'|^{-2n+1} \ .$$

In the first integral in the right–hand side of (II.6.12) we deform the domain of σ–integration from Γ to the image of Γ under the map $\sigma \rightarrow \mu(\zeta,s,\zeta',s',\sigma)\sigma$; this is permitted thanks to (II.6.1). We stress the fact that the image of Γ is contained in the cone $\mathcal{C}_\kappa \cap \mathcal{C}_1$. Once the deformation has been carried out, (II.6.10) & (II.6.13) entail that the norm of the integrand, in the first integral at the right in (II.6.12), does not exceed a constant times $(1+|\sigma|^2)^{-N}|\zeta-\zeta'|^{-2n+1}$. Choosing N sufficiently large ensures that the integral in question converges uniformly, as $\varepsilon \rightarrow +0$. Further exploiting property (II.6.13) leads to a similar conclusion for each one of the partial derivatives of the first integral at the right in (II.6.12): one must exploit the obvious identities

$$(\partial/\partial\zeta_i)|\zeta-\zeta'|^{-2n+1} = (-\partial/\partial\zeta_i')|\zeta-\zeta'|^{-2n+1} ,$$

$$(\partial/\partial\overline{\zeta}_i)|\zeta-\zeta'|^{-2n+1} = (-\partial/\partial\overline{\zeta}_i')|\zeta-\zeta'|^{-2n+1} ,$$

and use integration by parts in ζ' to transfer the derivatives to the other factors in the integrand. We leave the details to the reader.

In estimating the second integral in the right–hand side of (II.6.12) we avail ourselves of the following consequence of Cor. II.4.1 and (II.4.11):

$$s \in \partial\Theta \;\rightarrow\; \forall\; \zeta \in \Omega \;,\; \mathcal{Q}(s+i\psi(\zeta,s)) \equiv 0 \;.$$

If we take this into account in (II.5.8) we obtain that

$$s' \in \partial\Theta \;\rightarrow\; \widetilde{a}\,(\zeta,s,\zeta',s',\sigma,t) \equiv 0 \;.$$

It follows that the second integral at the right in (II.6.12) is equal to

$$\text{(II.6.14)} \quad \int_\Gamma \left\{ \int_{\Omega\times\partial\Theta} e^{i\sigma\cdot[s+i\psi(\zeta,s)-s'-i\psi(0,s')]-\varepsilon|\sigma|^2} \imath_N' \left[f\,(\zeta',s')\wedge E_{\widetilde{a}} \right] \right\} d\sigma \;.$$

But since $\psi(\zeta,s) = \varphi(H(\zeta,s),s)$ we have $\psi(0,s) = \varphi(0,s)$ and

(II.6.15) $|\psi(\zeta,s) - \varphi(0,s)| \leq const.|\mathcal{Q}(s+\imath\varphi(0,s))| \cdot |\zeta| ,$

thanks to (II.4.6).

Let $\varphi_1(s,s')$ denote the $d \times d$ matrix such that

$$\varphi(0,s) - \varphi(0,s') = \varphi_1(s,s')(s-s') ,$$

and ${}^t\varphi_1(s,s')$ its transpose. Observe that the norm of $\varphi_1(s,s')$ $(= I-J(0,s,s'))$ is very small in comparison to 1 . In (II.6.14) we deform the domain of σ–integration from Γ to the image of Γ under the map

$$\sigma \rightarrow \tilde{\sigma} = [I-\imath{}^t\varphi_1(s,s')][\sigma+\imath|\sigma|(s-s')] .$$

It is at this point that we avail ourselves of property (II.6.2). Combining it with (II.6.15) yields, for some suitably large constant $C > 0$ and all $\zeta \in \Omega$ and all $s \in \theta$, $s' \in \partial\theta$,

(II.6.16) $\mathcal{Im}\left[\tilde{\sigma}\cdot[s+\imath\psi(\zeta,s)-s'-\imath\psi(0,s')]\right] \geq$

$$\mathcal{Im}\left[\tilde{\sigma}\cdot[s+\imath\varphi(0,s)-s'-\imath\varphi(0,s')]\right] - C|\sigma||\zeta|(\mathrm{dist}(s,\partial\theta))^2.$$

On the other hand we have

$$\mathcal{Im}\left[\tilde{\sigma}\cdot[s+\imath\varphi(0,s)-s'-\imath\varphi(0,s')]\right] =$$

$$\mathcal{Im}\left[[\sigma+\imath|\sigma|(s-s')]\cdot[I-\imath\varphi_1(s,s')][s+\imath\varphi(0,s)-s'-\imath\varphi(0,s')]\right] =$$

$$|\sigma|(s-s')]\cdot[I+\varphi_1^2(s,s')](s-s')]\right] \geq |\sigma||s-s'|^2 .$$

But of course, $\mathrm{dist}(s, \partial\theta) \leq |s-s'|$ and thus, if Ω is chosen so small that $C|\zeta| < 1/2$, we obtain

$$\mathcal{I}m\left[\tilde{\sigma}\cdot[s+\imath\psi(\zeta,s)-s'-\imath\psi(0,s')]\right] \geq \tfrac{1}{2}|\sigma||s-s'|^2 .$$

As a consequence of this, if s stays in some compact subset K of θ, then the norm of the integrand in the integral (II.6.14) (after deformation of the domain of σ–integration) will not exceed $const.|z-z'|^{-2n+1}e^{-c|\sigma|}$ for some number $c > 0$ depending on K. As $\varepsilon \to +0$ the integral (II.6.14), and each one of its derivatives with respect to s', will converge uniformly. The same will be true of its partial derivatives of any order with respect to z and $\bar{z}$, as one sees by using integration by parts, as indicated earlier. $\bullet$

Under the hypotheses of Lemma II.6.2 we set, for any differential form $f \in C^\infty(\mathscr{C}\ell\mathcal{O};\Lambda^{m,q})$, when $\Gamma = \mathbb{R}^d\backslash\{0\}$,

$$\check{\mathrm{K}}_\mathcal{O}f = \lim_{\varepsilon\to+0} \check{\mathrm{K}}^\varepsilon_{\mathcal{O},\Gamma}f \ , \ \check{\mathrm{K}}_{\partial\Omega\times\theta}f = \lim_{\varepsilon\to+0} \check{\mathrm{K}}^\varepsilon_{\partial\Omega\times\theta,\Gamma}f \ ,$$

$$\check{\mathrm{B}}_{\partial\Omega\times\theta}f = \lim_{\varepsilon\to+0} \check{\mathrm{B}}^\varepsilon_{\partial\Omega\times\theta,\Gamma}f .$$

When $d = 1$ and $\Gamma = \mathbb{R}_\pm$, we write respectively $\check{\mathrm{K}}^\pm_\mathcal{O}$, $\check{\mathrm{K}}^\pm_{\partial\Omega\times\theta}$, $\check{\mathrm{B}}^\pm_{\partial\Omega\times\theta}$. We remind the reader that $\check{\mathrm{K}}_\mathcal{O}$, $\check{\mathrm{K}}^\pm_\mathcal{O}$, $\check{\mathrm{K}}_{\partial\Omega\times\theta}$, $\check{\mathrm{K}}^\pm_{\partial\Omega\times\theta}$ are defined by means of the maps ψ and $\tilde{a}\big|_{t=0}$, whereas $\check{\mathrm{B}}_{\partial\Omega\times\theta}$ & $\check{\mathrm{B}}^\pm_{\partial\Omega\times\theta}$ involve an integration with respect to t and thus require the definition of $\tilde{a}$ for all $t \in [0,1]$.

REMARK II.6.1.– The proof of Lemma II.6.2 shows that convergence of the operators $\mathrm{K}^\varepsilon_{\mathcal{O},\Gamma}$ & $\mathrm{K}^\varepsilon_{\partial\Omega\times\theta,\Gamma}$, and not just that of the operators $\check{\mathrm{K}}^\varepsilon_{\mathcal{O},\Gamma}$ & $\check{\mathrm{K}}^\varepsilon_{\partial\Omega\times\theta,\Gamma}$, takes place when the basic inequality, (II.6.6), is replaced by a stronger one, valid in the original domain, for instance the following:

(II.6.17)   *There is a map* (II.6.4), *endowed with property* (II.6.5), *and a*
constant $C_0 > 0$ *such that*

$$\mathscr{I}m\left[\mu(z,s,z',s',\sigma)\sigma\cdot[s+\imath\varphi(z,s)-s'-\imath\varphi(z',s')] \; - \right.$$

$$\left. \imath\,a\,(z,s,z',s',\mu(z,s,z',s',\sigma)\sigma)\cdot(z-z')\right] \; \geq \; |s-s'|^2/C_0 \; - \; C_0$$

$$\forall \; (z,s,z',s',\sigma) \in \mathcal{O}\times\mathcal{O}\times\Gamma \; .$$

In (II.6.17) there is no mention of t : this is because inequality (II.6.17) will be
used in connection with Corollaries II.2.3 & II.2.4, in which t does not enter. ❸

II.7. EXACT HOMOTOPY FORMULAS

As a conclusion to Ch. II we combine Theorems II.5.1, II.5.2, II.5.3 & II.5.4 with Lemmas II.6.1 & II.6.2. We shall assume, throughout and without recalling it, that diam Ω is sufficiently small to warrant the conclusions in Lemmas II.6.1 & II.6.2. In the statements that follow it will be assumed (but not recalled) that the maps a & μ are defined according to (II.5.2) & (II.6.4), and have properties (II.5.3) & (II.6.5) respectively. The map $\tilde{a}$, which enters in the definitions (at the end of Sect. II.6) of $\check{K}_{\mathcal{O}}$, $\check{K}_{\partial\Omega\times\theta}$ and $\check{B}_{\partial\Omega\times\theta}$, conforms to the requirements in (II.5.8). We set

$$ K_{\mathcal{O}} = \wp_*\check{K}_{\mathcal{O}}\wp^* \ , \ K_{\partial\Omega\times\theta} = \wp_*\check{K}_{\partial\Omega\times\theta}\wp^* \ , \ B_{\partial\Omega\times\theta} = \wp_*\check{B}_{\partial\Omega\times\theta}\wp^* \ . $$

THEOREM II.7.1.– *Assume* (II.6.6) *to be valid at* $t = 0$.

If (II.2.13) *holds in* $\hat{\mathcal{O}}\times\hat{\mathcal{O}}\times(\mathbb{R}^d\backslash\{0\})$ *then, for all differential forms* $f \in C^\infty(\mathscr{C}\ell\hat{\mathcal{O}}; \Lambda^{m,0})$ *we have, in* $\hat{\mathcal{O}}$,

(II.7.1) $(-1)^d f \ = \ - \ K_{\mathcal{O}}df \ + \ K_{\partial\Omega\times\theta}f$.

If (II.2.14) *holds in* $\hat{\mathcal{O}}\times\hat{\mathcal{O}}\times(\mathbb{R}^d\backslash\{0\})$ *then, for all differential forms* $f \in C^\infty(\mathscr{C}\ell\hat{\mathcal{O}}; \Lambda^{m,n})$ *we have, in* $\hat{\mathcal{O}}$,

(II.7.2) $f \ = \ (-1)^m d(K_{\mathcal{O}}f)$.

THEOREM II.7.2.– *Suppose* $1 \leq q \leq n-1$ *and* $\Gamma = \mathbb{R}^d\backslash\{0\}$. *Assume* (II.6.6) *to be valid at* $t = 0$; *also that* (II.6.11) *holds, and that the following is true:*

ı)   (II.5.10) *holds for all* $(z,s,z',s',\sigma) \in \hat{\mathcal{O}}\times\hat{\mathcal{O}}\times\Gamma$, *when* $t = 0$;

ıı)   (II.2.10) *holds for all* $(z,s,z',s',\sigma) \in \hat{\mathcal{O}}\times\hat{\mathcal{O}}\times\Gamma$, *when* $t = 1$;

ııı)   (II.5.11) *holds for all* $(z,s,z',s',\sigma) \in \hat{\mathcal{O}}\times\partial\hat{\mathcal{O}}\times\Gamma$, $s' \notin \partial\theta$, $0 \leq t \leq 1$.

Then, for all differential forms $f \in C^\infty(\mathscr{C}\ell\hat{\mathcal{O}};\Lambda^{m,q})$ *we have, in the open set* $\hat{\mathcal{O}}_0$

(see (II.6.11)),

$$(\text{II.7.3}) \qquad f \; = \; d[(-1)^q (K_{\mathcal{O}} - B_{\partial\Omega\times\theta}) f] \; + \; (-1)^{d+q+1} (K_{\mathcal{O}} - B_{\partial\Omega\times\theta}) df .$$

THEOREM II.7.3.– *Suppose* $1 \leq q \leq n{-}1$ *and* $\Gamma = \mathbb{R}^d \backslash \{0\}$. *Assume* (II.6.6) *to be valid at* $t = 0$; *assume furthermore that* (II.6.11) *holds, and that the following is true:*

  i)   (II.2.10) *holds for all* $(z,s,z',s',\sigma) \in \hat{\mathcal{O}} \times \hat{\mathcal{O}} \times \Gamma$, *when* $t = 1$;

  ii)   (II.5.12) *holds for all* $(z,s,z',s',\sigma) \in \hat{\mathcal{O}} \times \hat{\mathcal{O}} \times \Gamma$, $0 \leq t \leq 1$.

  Then, for all closed differential forms $f \in C^\infty(\mathscr{C}\!\ell\,\hat{\mathcal{O}}; \Lambda^{m,q})$ *we have, in the open set* $\hat{\mathcal{O}}_0$,

$$(\text{II.7.4}) \qquad\qquad f \; = \; d[(-1)^q (K_{\mathcal{O}} - B_{\partial\Omega\times\theta}) f] .$$

  Next we consider the case $d = 1$, $\Gamma = \mathbb{R}_\pm$. We define (see end of Sect. II.6)

$$I_\theta^\pm \; = \; \wp_* \hat{I}_\theta^\pm \wp^* , \; K_{\mathcal{O}}^\pm \; = \; \wp_* \hat{K}_{\mathcal{O}}^\pm \wp^* , \; K_{\partial\Omega\times\theta}^\pm \; = \; \wp_* \hat{K}_{\partial\Omega\times\theta}^\pm \wp^* , \; B_{\partial\Omega\times\theta}^\pm \; = \; \wp_* \hat{B}_{\partial\Omega\times\theta}^\pm \wp^* .$$

THEOREM II.7.4.– *Suppose* $d = 1$. *Assume* (II.6.6) *with* $\Gamma = \mathbb{R}_+$ *to be valid at* $t = 0$. *If* (II.2.13) *holds in* $\hat{\mathcal{O}} \times \hat{\mathcal{O}} \times \mathbb{R}_+$ *then, for all* $f \in C^\infty(\mathscr{C}\!\ell\,\hat{\mathcal{O}}; \Lambda^{n+1,0})$ *we have, in* $\hat{\mathcal{O}}$,

$$(\text{II.7.5}) \qquad\qquad I_\theta^+ f \; = \; K_{\mathcal{O}}^+ df \; - \; K_{\partial\Omega\times\theta}^+ f .$$

  If (II.2.14) *holds in* $\hat{\mathcal{O}} \times \hat{\mathcal{O}} \times \mathbb{R}_+$ *then, for all* $f \in C^\infty(\mathscr{C}\!\ell\,\hat{\mathcal{O}}; \Lambda^{n+1,n})$ *we have, in* $\hat{\mathcal{O}}$,

$$(\text{II.7.6}) \qquad\qquad I_\theta^+ f \; = \; (-1)^{n-1} d(K_{\mathcal{O}}^+ f) .$$

THEOREM II.7.5.– *Suppose* $d = 1$ *and* $1 \leq q \leq n{-}1$. *Assume* (II.6.6) *to be valid at* $t = 0$ *as well as* (II.6.11), *both with* $\Gamma = \mathbb{R}_+$, *and that the following is true:*

$\imath$)   (II.5.10) *holds for all* $(z,s,z',s',\sigma) \in \hat{O} \times \hat{O} \times \mathbb{R}_+$, *when* $t = 0$;

$\imath\imath$)   (II.2.10) *holds for all* $(z,s,z',s',\sigma) \in \hat{O} \times \hat{O} \times \mathbb{R}_+$, *when* $t = 1$;

$\imath\imath\imath$)   (II.5.11) *holds for all* $(z,s,z',s',\sigma) \in \hat{O} \times \partial\hat{O} \times \mathbb{R}_+$, $s' \notin \partial\Theta$, $0 \le t \le 1$.

Then, for all differential forms $f \in C^\infty(\mathscr{C}\ell\hat{O}; \Lambda^{n+1,q})$ *we have, in the open set* $\hat{O}_0$,

$$(\text{II}.7.7) \qquad \mathbf{I}_\theta^+ f = d[(-1)^q(\mathbf{K}_O^+ - \mathbf{B}_{\partial\Omega\times\theta}^+)f] + (-1)^q(\mathbf{K}_O^+ - \mathbf{B}_{\partial\Omega\times\theta}^+)df .$$

THEOREM II.7.6.– *Suppose* $d = 1$ *and* $1 \le q \le n{-}1$. *Assume* (II.6.6) *to be valid at* $t = 0$ *as well as* (II.6.11), *both with* $\Gamma = \mathbb{R}_+$, *and that the following is true:*

$\imath$)   (II.2.10) *holds for all* $(z,s,z',s',\sigma) \in \hat{O} \times \hat{O} \times \mathbb{R}_+$, *when* $t = 1$;

$\imath\imath$)   (II.5.12) *holds for all* $(z,s,z',s',\sigma) \in \hat{O} \times \hat{O} \times \mathbb{R}_+$, $0 \le t \le 1$.

Then, for all closed differential forms $f \in C^\infty(\mathscr{C}\ell\hat{O}; \Lambda^{n+1,q})$, *we have, in the open set* $\hat{O}_0$,

$$(\text{II}.7.8) \qquad\qquad \mathbf{I}_\theta^+ f = d[(-1)^q(\mathbf{K}_O - \mathbf{B}_{\partial\Omega\times\theta})f] .$$

Of course, the statements analogous to Theorems II.7.4, II.7.5, II.7.6, with subscripts and superscripts − instead of + , are true.

Actually we can also exploit Remark II.6.1. Suppose property (II.6.17) holds. In this case we define

$$\mathbf{K}_O = \lim_{\varepsilon \to +0} \mathbf{K}_{O,\Gamma}^\varepsilon \quad \text{when } \Gamma = \mathbb{R}^d\backslash\{0\} ,$$

$$\mathbf{I}_\theta^\pm = \lim_{\varepsilon \to +0} \mathbf{I}_{O,\Gamma}^\varepsilon , \ \mathbf{K}_O^\pm = \lim_{\varepsilon \to +0} \mathbf{K}_{O,\Gamma}^\varepsilon \quad \text{when } \Gamma = \mathbb{R}_\pm .$$

These operators are defined in the original CR structure (defined by means of the functions z_i and $w_k = s_k + \imath\varphi_k(z,s)$) on the set $\mathscr{D} \times \mathscr{B} \supset O$ (see Sect. II.3).

THEOREM II.7.7.– *Assume* (II.6.17) *to be valid.*

If (II.2.14) *holds in* $\mathcal{O} \times \mathcal{O} \times (\mathbb{R}^d \backslash \{0\})$ *then, for all* $f \in C^\infty(\mathcal{B}\mathcal{L}\mathcal{O}; \Lambda^{m,n})$ *we have, in* $\mathcal{O}$,

(II.7.9)
$$f = (-1)^m d(K_{\mathcal{O}} f) \;,$$

and when $d = 1$ & $\Gamma = \mathbb{R}_\pm$,

(II.7.10)
$$I^\pm_\theta f = (-1)^{n-1} d(K^\pm_{\mathcal{O}} f) \;.$$

It is pointless to try to exploit inequality (II.6.17) to prove the analogues of Theorems II.7.2 & II.7.5 (or, for that matter, II.7.3 & II.7.6) in the original set $\mathcal{O}$. In the proofs of those theorems the transformation $(\zeta, s) \to (H(\zeta, s), s)$ was needed not only to ensure the convergence of the homotopy operators but also to rid us of the integrals over the part $\Omega \times \partial\theta$ of the boundary of $\mathcal{O}$.

We close the chapter by a remark on the *microlocal* significance of homotopy formulas such as (II.7.7) or (II.7.8). Let us return to the proof of Lemma II.6.1 and consider first a form $f \in C^\infty(\mathcal{B}\mathcal{L}\mathcal{O}; \Lambda^{m,q})$. When $\Gamma = \mathbb{R}_-$ the integrals

$$\int_\Gamma \int_\theta e^{\imath \sigma \cdot [\tau - s' - \imath \psi(\zeta, s')] - \varepsilon \sigma \cdot \sigma} \, f_J(\zeta, s')[\det(I + \imath \psi_s(\zeta, s'))] ds' d\sigma$$

converge (in the C^∞ topology) in the region

$$\{ (\zeta, \tau) \in \mathbb{C}^{n+1} \;;\; (\zeta, \mathcal{R}e\,\tau) \in \mathcal{O} \;,\; \mathcal{I}m\,\tau < \psi(\zeta, \mathcal{R}e\,\tau) \} \;,$$

to a limit that is obviously holomorphic with respect to τ . When $\Gamma = \mathbb{R}_+$ the same is true in the analogous region with $\mathcal{I}m\,\tau > \psi(\zeta, s)$. The differential form $f - I^+_\theta f \in C^\infty(\mathcal{O}; \Lambda^{m,q})$ extends smoothly, and holomorphically with respect to τ , to the region $\mathcal{I}m\,\tau < \psi(\zeta, \mathcal{R}e\,\tau)$. If now $f \in C^\infty(\mathcal{B}\mathcal{L}\hat{\mathcal{O}}; \Lambda^{m,q})$ we obtain that $f - I^+_\theta f$

extends smoothly, and holomorphically with respect to τ , to the region $\mathcal{I}m\ w <$ $\varphi(z, \mathcal{R}e\ w)$. It is readily checked, by integration by parts, that when f is closed so is the extension: of course, this means that the latter is $\bar{\partial}$-closed.

Thus the right-hand side in (II.7.7) is equal to f modulo a C^{∞} form which is the C^{∞} boundary value of a differential form of bidegree (m,q) in the region that lies "below" the hypersurface Σ . When f is closed the right-hand side in (II.7.8) is equal to f modulo a C^{∞} form which is the C^{∞} boundary value of a $\bar{\partial}$-closed differential form of bidegree (m,q) in the region that lies "below" the hypersurface Σ .

CHAPTER III

GEOMETRIC CONDITIONS

III.1.   INVARIANCE OF THE CENTRAL HYPOTHESIS
IN THE HYPERSURFACE CASE

All the sections of the present chapter, except the last one, are devoted to the case $d = 1$. We denote by Σ the image of $\mathscr{D} \times \mathscr{B}$ under the map $(z,s) \to (z,w) \in \mathbb{C}^{n+1}$, with $w = s + i\varphi(z,s)$, $\varphi \in C^{\infty}(\mathscr{D} \times \mathscr{B})$ real–valued and $\varphi\big|_0 = 0$, $d\varphi\big|_0 = 0$. Thus Σ is the real hypersurface in $\mathbb{C}^{n+1}$ defined by the equation

(III.1.1) $$\mathscr{I}m\ w = \varphi(z, \mathscr{R}e\ w),$$

with $z \in \mathscr{D}$, $\mathscr{R}e\ w \in \mathscr{B}$.

We consider a biholomorphism of an open subset of $\mathbb{C}^{n+1}$ containing the hypersurface Σ, onto another open neighborhood of the origin in $\mathbb{C}^{n+1}$:

$$Z = F(z,w)\ ,\ W = G(z,w)\ ;$$

F & G are holomorphic and their Jacobian determinant, with respect to (z,w), does not vanish in $\eth$; $F\big|_0 = 0$, $G\big|_0 = 0$. The inverse of the mapping (F,G) will be denoted by

$$z = f(Z,W)\ ,\ w = g(Z,W)\ .$$

We want to express the fact that the submanifold Σ_1 which is the image of Σ under the mapping (F,G) is in the same position as Σ, i. e., Σ_1 is tangent to the real hyperplane $\mathscr{I}m\ W = 0$ at the origin. This means that Σ_1 is defined by an equation

$$\mathscr{I}m\ W = \Phi(Z, \mathscr{R}e\ W)\ ,$$

with Φ a real C^{∞} function in an appropriate domain in $\mathbb{C}^n \times \mathbb{R}^d$, $\Phi\big|_0 = 0$, $d\Phi\big|_0 = 0$.

Since the differential of the function

$$\mathcal{I}m\, g(Z, \mathcal{R}e\, W + \imath\Phi(Z, \mathcal{R}e\, W))$$

vanishes at the origin and since g is holomorphic we get

(III.1.2) $\partial g/\partial Z\big|_0 = 0$, $\partial(\mathcal{I}m\, g)/\partial(\mathcal{R}e\, W)\big|_0 = 0$.

Because the Jacobian matrices are nonsingular we conclude that

(III.1.3) $\det(\partial f\,/\partial Z)\big|_0 \neq 0$,

and that $\partial(\mathcal{R}e\, g)/\partial(\mathcal{R}e\, W)\big|_0 \neq 0$. We shall further assume that the transformation (F,G) preserves the meaning of "above" and "below" the hypersurface under consideration, i. e., transforms the region $\mathcal{I}m\, w > \varphi(z, \mathcal{R}e\, w)$ into the region $\mathcal{I}m\, W > \varphi(Z, \mathcal{R}e\, W)$. This is equivalent to assuming

(III.1.4) $\partial g/\partial(\mathcal{R}e\, W)\big|_0 > 0$.

We are then going to hypothesize a strong version of the central condition (II. 6.6); the version we consider is the one that obtains in the forthcoming Sect. III.3. We shall take $\Gamma = \mathbb{R}_+$; the case $\Gamma = \mathbb{R}_-$ is treated in similar fashion. We are going to assume that the map $\sigma^{-1}a$ and μ are independent of $\sigma > 0$. Under these circumtances, after we divide both sides in (II.6.6) by σ and let σ go to $+\infty$, we get that (II.6.6) is valid for $\sigma = 1$ and $C_0 = 0$. After multiplying back by σ we recover (II. 6.6) with $C_0 = 0$. Here we regard t as a parameter, and we omit it from further consideration.

In summary we assume that there is a C^∞ map $a : (\mathcal{D} \times \mathcal{B})\times(\mathcal{D} \times \mathcal{B}) \to \mathbb{C}^n$, and a complex–valued function μ in $(\mathcal{D} \times \mathcal{B})\times(\mathcal{D} \times \mathcal{B})$ conforming to the requirements in (II.6.5) and such that

(III.1.5) $\mathcal{I}m\left[\mu[w-w'-\imath a\cdot(z-z')]\right] \geq 0$.

We have used the notation $w = s+\imath\varphi(z,s)$, $w' = s'+\imath\varphi(z',s')$.

LEMMA III.1.1.– *If* (III.1.5) *holds then necessarily* $a = 0$ *if* $z = z' = 0$, $s = s' = 0$.

$\mathcal{P}roof$: When $z' = 0$ & $s = s' = 0$ the function at the right, in (III.1.5), is equal to

$$\mathcal{R}e\left[\mu[\int_0^1 \varphi_z(\lambda z,0)\mathrm{d}\lambda + a\ ]\cdot z\right]\ ,$$

which is everywhere ≥ 0 and vanishes at the origin. Its differential must vanish at the origin. Since μ cannot vanish, by (II.6.3), and $\varphi_z\big|_0 = 0$ necessarily $a\big|_0 = 0$.∎

Setting $z = f(Z,W)$, $w = g(Z,W)$ yields

(III.1.6)

$$w - w' - \imath a\cdot(z-z') = (g_2 - \imath a\cdot f_2)(W - W') + \imath({}^t f_1 a - \imath g_1)\cdot(Z - Z')\ .$$

Keep in mind that, provided diam $(\mathcal{D}\times\mathcal{B})$ is small enough, the $n\times n$ matrix f_1 (whose transpose we have denoted by ${}^t f_1$) is as close as we wish to the Jacobian matrix $\partial f/\partial Z$, the n–vectors f_2 & g_1 are arbitrarily close to $\partial f/\partial(\mathcal{R}e\ W)$ and to $\partial g/\partial Z$ respectively, while the scalar g_2 is arbitrarily close to $\partial g/\partial(\mathcal{R}e\ W)$. It follows from (III.1.2), (III.1.3) & (III.1.4) that there is a constant $c > 0$ such that, given any $\varepsilon > 0$, if diam $(\mathcal{D}\times\mathcal{B})$ is sufficiently small, then

(III.1.7) $c + \varepsilon^{-1}\mathcal{I}m\ (g_2 - \imath a\cdot f_2) \leq \mathcal{R}e\ (g_2 - \imath a\cdot f_2)$.

This allows us to define

$$(III.1.8) \qquad M = (g_2 - \imath a \cdot f_2)\mu \ , \ A = (g_2 - \imath a \cdot f_2)^{-1}({}^t f_1 a - \imath g_1) \ ;$$

we have, by (III.1.6),

$$(III.1.9) \qquad \mu[w - w' - \imath a \cdot (z - z')] = M[W - W' - \imath A \cdot (Z - Z')] \ ,$$

and therefore condition (III.1.5) is also valid after we have carried out the biholo-morphism (F, G).

It follows from (II.6.5) and (III.1.6) that M also satisfies (II.6.5), in a suitable domain in $(Z, \mathscr{R}e\ W, Z', \mathscr{R}e\ W')$–space. On the other hand, due to the fact that f & g are holomorphic functions of (Z, W) , if the map a satisfies any condition of the kind (II.1.10) (or (II.1.11)) the same will be true of A – in the CR structure defined by the functions Z_i $(1 \leq i \leq n)$ & $W = S + \imath \Phi(Z, S)$.

This completes the proof of the invariance, under biholomorphic change of variables, of condition (III.1.5).

III.2.   THE HYPERSURFACE CASE : SUPPORTING MANIFOLDS

We continue to denote by Σ the real hypersurface in $\mathbb{C}^{n+1}$ defined by the equation (III.1.1). We introduce the following

DEFINITION III.2.1.- *Let ν be an integer, $0 \leq \nu \leq n$. We say that the hypersurface Σ* admits supporting manifolds of holomorphic type ν *if there is a C^{∞} map $\mathscr{D} \times \Sigma \ni$* $(z,\wp) \rightarrow A(z,\wp) \in \mathbb{C}^n$ *endowed with the following properties:*

(III.2.1)   *As $(z,\wp)$ ranges over $\mathscr{D} \times \Sigma$, the function*

$$\mathscr{I}m\, w(\wp) \;+\; \mathscr{R}e\,[A(z,\wp)\cdot(z-z(\wp))] \;-\; \varphi(z, \mathscr{R}e\, w(\wp)-\mathscr{I}m\,[A(z,\wp)\cdot(z-z(\wp))])$$

does not change sign;

(III.2.2) $\forall\, \wp \in \Sigma$, rank $(\partial A_i/\partial \bar{z}_j)_{1 \leq i, j \leq n} \leq n - \nu$.

If Σ admits supporting manifolds of holomorphic type ν it automatically admits supporting manifolds of any holomorphic type $\leq \nu$.

Let the map A satisfy the requirements in Def. III.2.1. As $\wp \in \Sigma$ is fixed and z ranges over $\mathscr{D}$, the point $(z, w(\wp)+\imath A(z,\wp)\cdot[z-z(\wp)])$ describes a $2n$-dimensional submanifold $\mathscr{M}(\wp)$ of $\mathbb{C}^{n+1}$ which passes through $\wp$ and which, by (III.2.1), stays on one side of Σ ; the side which contains $\mathscr{M}(\wp)$ is the same for all $\wp \in \Sigma$. We can regard Σ as the base in a fibre bundle whose fibres are the real submanifolds $\mathscr{M}(\wp)$. It is not precluded, however, that $\mathscr{M}(\wp)$ & $\mathscr{M}(\wp_*)$, $\wp \neq \wp_*$, intersect.

It is conveniemt to make the terminology a bit more precise, by saying that Σ admits supporting manifolds *from below* (resp., *from above*), of holomorphic type ν , if the map A in Def. III.2.1 satisfies condition (III.2.2), as well as the following estimate, for all $(z,\wp) \in \mathscr{D} \times \Sigma$,

(III.2.1')$_+$

$$\mathcal{I}m\ w(\wp)\ +\ \mathcal{R}e\,[A(z,\wp)\cdot(z{-}z(\wp))]\ -\ \varphi(z,\mathcal{R}e\ w(\wp){-}\mathcal{I}m\,[A(z,\wp)\cdot(z{-}z(\wp))])\ \leq\ 0$$

$$\Bigg[\text{resp.,}$$

(III.2.1')$_-$

$$\mathcal{I}m\ w(\wp)\ +\ \mathcal{R}e\,[A(z,\wp)\cdot(z{-}z(\wp))]\ -\ \varphi(z,\mathcal{R}e\ w(\wp){-}\mathcal{I}m\,[A(z,\wp)\cdot(z{-}z(\wp))])\ \geq\ 0\ \Bigg]\ .$$

It should be pointed out, however, that these notions of "above" and "below" depend on the embedding (just as does the number of positive, or the number of negative, eigenvalues of the Levi form).

PROPOSITION III.2.1.– *In order that, possibly after contracting $\mathscr{D}$ and $\mathscr{B}$ about the origin, Σ admit supporting manifolds from below (resp., above), of holomorphic type n, it is necessary and sufficient that there be an open neighborhood N of the origin in $\mathbb{C}^n$ and a C^∞ map $\gamma: N{\times}\Sigma \to \mathbb{C}^{n+1}$ endowed with the following properties:*

(III.2.3) $\forall\ \wp \in \Sigma$, $\gamma(0,\wp) = \wp$;

(III.2.4) $\forall\ \wp \in \Sigma$, *the map $\zeta \to \gamma(\zeta,\wp)$ is holomorphic and the Jacobian matrix $\partial(z{\circ}\gamma)/\partial\zeta$ is nonsingular;*

(III.2.5)   *in $N{\times}\Sigma$, $\mathcal{I}m\ w{\circ}\gamma\ -\ \varphi(z{\circ}\gamma,\mathcal{R}e\ w{\circ}\gamma) \leq 0$ (resp., ≥ 0).*

Proof: I. *Necessity.* Let A denote a map as in Def. III.2.1, satisfying (III.2.1') and (III.2.2) with $\nu = n$. Since rank $\bar\partial A \equiv 0$, the map A must be holomorphic with respect to z ; it suffices to take

$$\gamma(\zeta,\wp)\ =\ (z(\wp){+}\zeta,w(\wp)\ +\ \imath A(z(\wp){+}\zeta,\wp)\cdot\zeta)\ ,$$

and to contract the domain $\mathscr{D}$ and select the neighborhood N in such a way that

$z = z(\wp) + \zeta$ stays in the domains of definition of A and of $\varphi(\cdot, s)$ $(s \in \mathcal{B})$.

II. *Sufficiency.* Let γ be a map as in Prop. III.2.1. We can solve with respect to ζ the equation $z = z(\gamma(\zeta,\wp))$; the solution is a C^1 function of $(z,\wp)$ in $\mathcal{D} \times \mathcal{B}$ (suitably contracted about the origin), holomorphic with respect to z and equal to zero when $z = z(\wp)$. We set $w(\gamma(\zeta,\wp)) = H(z,\wp)$, with H smooth and holomorphic with respect to z . By (III.2.3), $H(z,\wp) = w(\wp) + \imath A(z,\wp) \cdot [z - z(\wp)]$, whence (III.2.1'), thanks to (III.2.5). ●

EXAMPLE III.2.1.– Suppose the hypersurface Σ is *Levi flat;* this is equivalent to assuming that Σ is foliated by holomorphic manifolds of complex dimension n . In this case Σ admits supporting manifolds of holomorphic type n both from below and from above, namely the leaves in the holomorphic foliation of Σ . Locally, Σ is biholomorphically equivalent to $\mathbb{C}^n \times \mathbb{R} \subset \mathbb{C}^{n+1}$. Recall that a complex hypersurface in $\mathbb{C}^{n+1}$ that contains the origin must either intersect both sides of $\mathbb{C}^n \times \mathbb{R}$ or else lie entirely in $\mathbb{C}^n \times \mathbb{R}$, *i. e.*, be identical to $\mathbb{C}^n \times \{0\}$. This means that, when Σ is Levi flat, the only supporting manifolds of holomorphic type n are the leaves of the holomor– phic foliation of Σ . ●

DEFINITION III.2.2.– *Let ν be an integer, $0 \leq \nu \leq n$. We say that the hypersurface Σ admits* strictly supporting *manifolds of holomorphic type ν if there is a C^∞ map from $\mathcal{D} \times \Sigma$ into $\mathbb{C}^n$, $(z,\wp) \to A(z,\wp)$, which satisfies condition (III.2.2) as well as the following condition:*

(III.2.6)   *As $(z,\wp)$ ranges over $\mathcal{D} \times \Sigma$, the function*

$$\mathcal{I}m\, w(\wp) + \mathcal{R}e\, [A(z,\wp) \cdot (z - z(\wp))] - \varphi(z, \mathcal{R}e\, w(\wp) - \mathcal{I}m\, [A(z,\wp) \cdot (z - z(\wp))])$$

does not change sign and does not vanish except when $z = z(\wp)$.

Evidently, if Σ admits strictly supporting manifolds of holomorphic type ν it admits **supporting** manifolds of holomorphic type ν ; each manifold $\mathcal{M}(\wp)$ (see above) stays on one side of Σ and intersects (and is tangent to) Σ only at the point $\wp$. Any Levi flat hypersurface admits supporting manifolds of holomorphic type n but does not admit strictly supporting manifolds of holomorphic type n (see Example III.2.1). It would be easy to construct similar examples for holomorphic types $\nu < n$.

We say that Σ admits strictly supporting manifolds *from below* (resp., *from above*) of holomorphic type ν , if the map A in Def. III.2.2, in addition to (III.2.2), satisfies, for all $(z,\wp) \in \mathcal{D} \times \Sigma$ such that $z \neq z(\wp)$, the inequality

(III.2.6')$_+$

$$\mathcal{I}m\, w(\wp) \; + \; \mathcal{R}e\,[A(z,\wp)\cdot(z-z(\wp))] \; - \; \varphi(z,\mathcal{R}e\, w(\wp)-\mathcal{I}m\,[A(z,\wp)\cdot(z-z(\wp))]) \; < \; 0$$

$\Bigg[$resp.,

(III.2.6')$_-$

$$\mathcal{I}m\, w(\wp) \; + \; \mathcal{R}e\,[A(z,\wp)\cdot(z-z(\wp))] \; - \; \varphi(z,\mathcal{R}e\, w(\wp)-\mathcal{I}m\,[A(z,\wp)\cdot(z-z(\wp))]) \; > \; 0 \Bigg]\,.$$

It is an elementary fact that one can simplify the Taylor expansion of φ at the origin to third order, by substituting

$$\tilde{w} \; = \; w \; - \; a_0 w^2 \; - \; \sum_{i=1}^{n} a_i z_i w \; - \; \sum_{j,k=1}^{n} b_{jk} z_j z_k$$

for w . The coefficients a_i $(i = 0,1,...,n)$, b_{jk} $(1 \leq j$, $k \leq n)$ can be chosen so as to ensure

$$\tilde{s} \; = \; \mathcal{R}e\, \tilde{w} \; = \; s \; + \; O(|z|^2+s^2)\,,$$

$$\mathscr{J}m\ \tilde{w} = \sum_{j,k=1}^{n} c_{jk} z_j \bar{z}_k + O(|z|^3 + |\tilde{s}|^3) \ ,$$

with $c_{jk} = \bar{c}_{jk}$. After deleting the tildas we may therefore assume, from the start, that

$$\varphi(z,s) = \sum_{j,k=1}^{n} c_{jk} z_j \bar{z}_k + O(|z|^3 + |s|^3) \ .$$

The self–adjoint $n{\times}n$ matrix $\varphi_{z\bar{z}}(0,0) = (c_{jk})_{1\leq j,k\leq n}$ is often thought of as the Levi form (see Sect. II.1) of the hypersurface Σ at the origin (note however that this notion is strongly dependent on the embedding). We shall call ν_+ (resp., ν_-) the number of eigenvalues > 0 (resp. < 0) of matrix $\varphi_{z\bar{z}}(0,0)$. When both ν_+ and ν_- are equal to zero it simply means that φ vanishes to third order at the origin. After a $\mathbb{C}$–linear change of variables we may assume

(III.2.7) $$\varphi(z,s) = |z_{(1)}|^2 - |z_{(2)}|^2 + \psi(z,s) \ ,$$

(III.2.8) $$|\psi(z,s)| \leq const.(|z|^3 + |s|^3) \ ,$$

where $z_{(1)} = (z_1, ..., z_{\nu_+})$, $z_{(2)} = (z_{\nu_+ +1}, ..., z_\nu)$, $\nu = \nu_+ + \nu_-$. We shall also write $z_{(3)} = (z_{\nu+1}, ..., z_n)$. Note, for use below, that $\varphi_s = \psi_s$.

PROPOSITION III.2.2.– *The hypersurface Σ admits strictly supporting manifolds from below of holomorphic type ν_+ and from above, of holomorphic type ν_- .*

Proof: Let p be an arbitrary point in Σ ; in the sequel we write $z' = z(p)$, $w' = w(p)$, $s' = \mathscr{R}e\ w(p)$; thus p is completely determined by (z',s') , which allows us to write $A(z,z',s')$ rather than $A(z,p)$. We may also regard (z,s,z',s') as the variable point in $(\mathscr{D} \times \mathscr{B}) \times (\mathscr{D} \times \mathscr{B})$.

Define $w = w' + \imath A \cdot (z-z')$; we have:

$$\varphi(z, \mathcal{R}e\; w) - \mathcal{I}m\; w = |z_{(1)}|^2 - |z_{(2)}|^2 + \psi(z, s' - \mathcal{I}m\;[A\cdot(z-z')]) -$$

$$\left[|z'_{(1)}|^2 - |z'_{(2)}|^2 + \psi(z', s') + \mathcal{R}e\;[A\cdot(z-z')]\right] =$$

$$|z-z'|^2 - \mathcal{R}e\;[A\cdot(z-z')] + 2\mathcal{R}e\;[\bar{z}'_{(1)}\cdot(z_{(1)}-z'_{(1)}) - \bar{z}_{(2)}\cdot(z_{(2)}-z'_{(2)})] -$$

$$|z_{(3)}-z'_{(3)}|^2 + \psi(z,s') - \psi(z',s') - \psi_s(z',s')\mathcal{I}m\;[A\cdot(z-z')] + O(\delta|z-z'|^2) =$$

$$|z-z'|^2 - \mathcal{R}e\left[[(1-\imath\psi_s(z',s'))A - 2\psi_z(z',s')]\cdot(z-z')\right] +$$

$$2\mathcal{R}e\;[\bar{z}'_{(1)}\cdot(z_{(1)}-z'_{(1)}) - \bar{z}_{(2)}\cdot(z_{(2)}-z'_{(2)})] - |z_{(3)}-z'_{(3)}|^2 + O(\delta|z-z'|^2) ,$$

where we have used the notation $\delta = \mathrm{diam}(\mathscr{D}\times\mathscr{B})$. If we select

$$\text{(III.2.9)}\qquad A = [1-\imath\varphi_s(z',s')]^{-1}\left[2\psi_z(z',s') + (2\bar{z}'_{(1)}, -2\bar{z}'_{(2)}, \bar{z}'_{(3)} - \bar{z}_{(3)})\right] ,$$

and if $\mathrm{diam}(\mathscr{D}\times\mathscr{B})$ is small enough we get

$$\text{(III.2.10)}\qquad\qquad \varphi(z, \mathcal{R}e\; w) - \mathcal{I}m\; w \geq \tfrac{1}{2}|z-z'|^2,$$

whence (III.2.6')_+ . The map $A : \mathbf{C}^n \times \Sigma \to \mathbf{C}^n$ is holomorphic with respect to $z_{(1)}$ and therefore satisfies condition (III.2.2) with $\nu = \nu_+$. In conclusion, Σ admits strictly supporting manifolds from below, of holomorphic type ν_+ .

Likewise, choosing

$$\text{(III.2.11)}\qquad A = [1-\imath\varphi_s(z',s')]^{-1}\left[2\psi_z(z',s') - (2\bar{z}'_{(1)}, -2\bar{z}'_{(2)}, \bar{z}'_{(3)} - \bar{z}_{(3)})\right]$$

leads to

(III.2.12) $$\varphi(z, \mathscr{R}e\, w) - \mathscr{I}m\, w \leq -\tfrac{1}{2}|z-z'|^2$$

provided $\text{diam}(\mathscr{D} \times \mathscr{B})$ is small enough. The map (III.2.11) is holomorphic with respect to $z_{(2)}$ and thus satisfies (III.2.2) with $\nu = \nu_-$: Σ admits strictly supporting manifolds from above, of holomorphic type ν_- . ●

COROLLARY III.3.1.– *The hypersurface Σ admits strictly supporting manifolds of holomorphic type zero.*

That a hypersurface admits supporting manifolds of a given type might not be determined by the eigenvalues of the Levi form at the origin, as shown in the following

EXAMPLE III.2.2.– We continue to give to $z_{(1)}$, $z_{(2)}$, $z_{(3)}$ the same meaning as in the preceding argument. Suppose

(III.2.13) $$\varphi = \varphi_1(z_{(1)}) - \varphi_2(z_{(2)}) \ .$$

To satisfy condition (III.2.1') we seek a map $A = (A_1, A_2, A_3)$ such that

(III.2.14)
$$\varphi_1(z_{(1)}) - \varphi_1(z'_{(1)}) - \mathscr{R}e\, A_1 \cdot (z_{(1)} - z'_{(1)}) \geq 0 \ ,$$

$$\varphi_2(z'_{(2)}) - \varphi_2(z_{(2)}) - \mathscr{R}e\, A_2 \cdot (z_{(2)} - z'_{(2)}) \geq 0 \ .$$

If we assume that

(III.2.15) $$\varphi_1 \ , \ \varphi_2 \ \text{are convex,}$$

we can select

$$A_1 = 2(\partial\varphi_1/\partial z_{(1)})(z'_{(1)}) \ , \ A_2 = -\ 2(\partial\varphi_2/\partial z_{(2)})(z_{(2)}) \ ,$$

(III.2.16)

$$A_3 = -\ (\bar{z}_{(3)} - \bar{z}'_{(3)}) \ ;$$

A is holomorphic with respect to $z_{(1)}$ and thus Σ admits supporting manifolds from below, of holomorphic type ν_+ . If instead we choose

$$A_1 = -\ 2(\partial\varphi_1/\partial z_{(1)})(z_{(1)}) \ , \ A_2 = -\ 2(\partial\varphi_2/\partial z_{(2)})(z'_{(2)}) \ ,$$

(III.2.17)

$$A_3 = \bar{z}_{(3)} - \bar{z}'_{(3)} \ ,$$

we see that Σ admits supporting manifolds from above, of holomorphic type ν_- .

If instead of the hypothesis (III.2.15) we require that φ_1 & φ_2 be *strictly convex* then, with the same choice (III.2.16), the inequalities (III.2.14) will be strict except when $z_{(1)} = z'_{(1)}$ or $z_{(2)} = z'_{(2)}$. In this case Σ admits strictly supporting manifolds from below, of holomorphic type ν_+ ; the choice (III.2.17) shows that Σ admits strictly supporting manifolds from above, of holomorphic type ν_- . ❷

Lastly we present a result of J. A. Petersen which shows that the standard notion of supporting (complex) manifolds is a particular case of Def. III.2.2.

DEFINITION III.2.3.– *We shall say that Σ admits* strictly supporting holomorphic *manifolds from below, of dimension ν , if there is an open polydisk Δ_1 in $\mathbb{C}^\nu$, centered at the origin, and a C^∞ map $\chi : \Delta_1 \times \Sigma \to \mathbb{C}^{n+1}$, holomorphic with respect to the variable ζ in Δ_1 and having the following properties:*

(III.2.18) $\forall \ \wp \in \Sigma \ , \ \chi(0,\wp) = \wp$;

(III.2.19)   rank $\partial\chi/\partial\zeta = \nu$;

(III.2.20)   *Jm $w < \varphi(z, \mathcal{R}e\ w)$ on the range of χ in $\Delta_1\backslash\{0\}$.*

PROPOSITION III.2.3.— *If Σ admits strictly supporting holomorphic manifolds from below, of dimension ν , then, possibly after shrinking Σ about the origin, it admits strictly supporting manifolds from below, of holomorphic type ν .*

Proof : Because of (III.2.20) we have necessarily $\partial_\zeta(w \circ \chi) = 0$ when $\zeta = 0$. It follows from (III.2.19) that the rank of $\partial_\zeta(z \circ \chi)$ must be equal to ν . After relabeling the variables (and contracting both Δ_1 & Σ about 0) we may assume that the Jacobian matrix $\partial_\zeta(z_{(1)} \circ \chi)$ is nonsingular. [We shall be using throughout the notation $z_{(1)} = (z_1 ,..., z_\nu)$, $z_{(2)} = (z_{\nu+1} ,..., z_n)$.] We may therefore substitute $z_{(1)} - z_{(1)}(\wp)$ for ζ and simply assume that we have a C^∞ map, holomorphic with respect to $z_{(1)}$,

$$(III.2.21) \qquad \Delta_1 \times \Sigma \ni (z_{(1)},\wp) \to (Z_{(2)}(z_{(1)},\wp), W(z_{(1)},\wp)) \in \mathbb{C}^{n-\nu+1},$$

such that, for all $\wp \in \Sigma$,

$$(III.2.22) \qquad Z_{(2)}(z_{(1)}(\wp),\wp) = z_{(2)}(\wp) \ , \ W(z_{(1)}(\wp),\wp) = w(\wp) \ ,$$

$$(III.2.23) \qquad \mathfrak{Im} \ W(z_{(1)},\wp) < \varphi(z_{(1)}, Z_{(2)}(z_{(1)},\wp), \ \mathfrak{Re} \ W(z_{(1)},\wp))$$
$$if \ z_{(1)} \neq z_{(1)}(\wp) \ .$$

Next we select, passing through any $\wp \in \Sigma$, a $(2n-2\nu)$–dimensional submanifold of Σ , $z_{(2)} \to \mathcal{P}(z_{(2)},\wp)$, transversal to the holomorphic submanifold of $\mathbb{C}^{n+1}$ described by $(z_{(1)}, Z_{(2)}(z_{(1)},\wp), W(z_{(1)},\wp))$. We take

$$\mathcal{P}(z_{(2)},\wp) = (z_{(1)}(\wp), z_{(2)}, W^*(z_{(2)},\wp)) \ , \ W^*(z_{(2)},\wp) = s^* + \imath\varphi(z_{(1)}(\wp), z_{(2)}, s^*) \ ,$$

$$s^* = s(\wp) + \mathfrak{Re} \ \{\theta \cdot [z_{(2)} - z_{(2)}(\wp)] + \Gamma[z_{(2)} - z_{(2)}(\wp)] \cdot [z_{(2)} - z_{(2)}(\wp)]\}$$

with $\theta \in \mathbb{C}^{n-\nu}$ and Γ a complex $(n-\nu) \times (n-\nu)$ matrix; θ & Γ will be chosen below.

For all $p \in \Sigma$,

(III.2.24) $$\mathcal{P}(z_{(2)},p) \in \Sigma , \; \mathcal{P}(z_{(2)}(p),p) = p .$$

Note that

(III.2.25) $$Z_{(2)}(z_{(1)}(p),\mathcal{P}(z_{(2)},p)) = z_{(2)}(\mathcal{P}(z_{(2)},p)) = z_{(2)} .$$

This has the following immediate consequence, if Σ is suitably contracted about 0 :

(III.2.26)   *the map $z \rightarrow u$, with $u_{(1)} = z_{(1)}$, $u_{(2)} = Z_{(2)}(z_{(1)},\mathcal{P}(z_{(2)},p))$ is a C^∞ diffeomorphism, holomorphic with respect to $z_{(1)}$, of an open polydisk Δ in $\mathbb{C}^n$, centered at 0 , onto a open neighborhood of 0 in $\mathbb{C}^n$; moreover, $z = z(p)$ if and only if $u = z(p)$.*

We define

$$W^{\#}(z,p) = W(z_{(1)},\mathcal{P}(z_{(2)},p)) - \imath |z_{(2)}-z_{(2)}(p)|^2 .$$

By (III.2.22) $W^{\#}(z(p),p) = p$; $W^{\#}(z,p)$ is holomorphic with respect to $z_{(1)}$. Since we can contract Σ about 0 to achieve $|\varphi_s| << 1$ we derive from (III.2.23):

(III.2.27) $$\mathcal{I}m \; W^{\#}(z,p) < \varphi(z_{(1)},Z_{(2)}(z_{(1)},p),\mathcal{R}e \; W^{\#}(z,p)) -$$

$$\tfrac{1}{2}|z_{(2)}-z_{(2)}(p)|^2 \;\; if \; z_{(1)} \neq z_{(1)}(p) .$$

At this stage we select θ & Γ in the definition of $\tilde{s}$ to ensure

(III.2.28) $$(\partial/\partial\bar{z}_{(2)}) W^{\#}(z,p)\big|_{z=z(p)} = 0 \; , \; (\partial/\partial\bar{z}_{(2)})^2 W^{\#}(z,p)\big|_{z=z(p)} = 0 .$$

Write $\Delta = \Delta_1 \times \Delta_2$, with $\Delta_1 \subset \mathbb{C}^\nu$, $\Delta_2 \subset \mathbb{C}^{n-\nu}$. Property (III.2.28) allows us to write

$$W^{\#}(z_{(1)}(\wp), z_{(2)}, \wp) = W^{\#}(z(\wp), \wp) - \imath A_{(2)}(z_{(2)}, \wp) \cdot (z_{(2)} - z_{(2)}(\wp)) +$$

$$O(|z_{(2)} - z_{(2)}(\wp)|^3) \ .$$

We may combine this with the following consequence of (III.2.25):

$$W^{\#}(z, \wp) + \imath A_{(2)}(z_{(2)}, \wp) \cdot [Z_{(2)}(z_{(1)}, \mathcal{P}(z_{(2)}, \wp)) - z_{(2)}(\wp)] =$$

$$W^{\#}(z_{(1)}(\wp), z_{(2)}, \wp) + \imath A_{(2)}(z_{(2)}, \wp) \cdot (z_{(2)} - z_{(2)}(\wp)) - \imath A_{(1)}(z, \wp) \cdot (z_{(1)} - z_{(1)}(\wp)) \ ,$$

where $A_{(1)}(z, \wp)$ is holomorphic with respect to $z_{(1)}$. We get

$$W^{\#}(z, \wp) = w(\wp) - \imath A(u, \wp) \cdot (u - z(\wp)) + O(|z_{(2)} - z_{(2)}(\wp)|^3) \ .$$

We have used the notation (*cf.* (III.2.26))

$$A(u, \wp) = (A_{(1)}(z, \wp), A_{(2)}(z_{(2)}, \wp)).$$

We put this into (III.2.27); requiring that $|z_{(2)} - z_{(2)}(\wp)|$ be suitably small gets us

$$\mathscr{I}m \left[w(\wp) - \imath A(u, \wp) \cdot (u - z(\wp)) \right] <$$

$$\varphi(u, \mathscr{R}e \left[w(\wp) - \imath A(u, \wp) \cdot (u - z(\wp)) \right]) - \tfrac{1}{2} |z_{(2)} - z_{(2)}(\wp)|^2 \quad if \ z_{(1)} \neq z_{(1)}(\wp) \ .$$

Using once again (III.2.26) we conclude that

$$\mathscr{I}m \left[w(\wp) - \imath A(u, \wp) \cdot (u - z(\wp)) \right] < \quad \varphi(u, \mathscr{R}e \left[w(\wp) - \imath A(u, \wp) \cdot (u - z(\wp)) \right]) \quad if \ u \neq z(\wp) \ ,$$

which entails (III.2.6). Finally, we note that

$$\partial A/\partial \overline{u} = (\partial z_{(2)}/\partial \overline{u})\partial A/\partial z_{(2)} + (\partial \overline{z}_{(2)}/\partial \overline{u})\partial A/\partial \overline{z}_{(2)} \, ,$$

$$0 = (\partial u_{(2)}/\partial z_{(2)})\partial z_{(2)}/\partial \overline{u} + (\partial A/\partial \overline{z}_{(2)})\partial \overline{z}_{(2)}/\partial \overline{u} \, ,$$

and $\partial u_{(2)}/\partial z_{(2)}$ is nonsingular. This shows that $\partial A/\partial \overline{u}$ is a multiple of $\partial \overline{z}_{(2)}/\partial \overline{u}$ whose rank is equal to $n - \nu$ and thus proves that rank $(\partial A/\partial \overline{u}) \leq n - \nu$. ▩

III.3.   LOCAL HOMOTOPY FORMULAS ON A HYPERSURFACE

Same notation as in the previous section. In particular we write $z' = z(\wp)$, w'
$= w(\wp)$, $s' = \mathscr{R}e\, w'$.

PROPOSITION III.3.1.– *Suppose the map A satisfies condition* (III.2.1')$_+$ *(resp.,* (III.2.
1')$_-$*) and set*

$$\beta(z,s,z',s') = -\int_0^1 \varphi_s(z,s+\lambda(s'-s-\mathscr{I}m\,[A(z,z',s')\cdot(z-z')]))\mathrm{d}\lambda\ .$$

Then, for all $(z,s,z',s') \in (\mathscr{D} \times \mathscr{B})\times(\mathscr{D} \times \mathscr{B})$,

(III.3.1)$_+$
$$\mathscr{I}m\,\left[[1+\imath\beta(z,s,z',s')][s+\imath\varphi(z,s)-s'-\imath\varphi(z',s')-\imath A(z,z',s')\cdot(z-z')]\right] \geq 0$$

$\left[resp.,\right.$
(III.3.1)$_-$
$$\mathscr{I}m\,\left[[1+\imath\beta(z,s,z',s')][s+\imath\varphi(z,s)-s'-\imath\varphi(z',s')-\imath A(z,z',s')\cdot(z-z')]\right] \leq 0\ \left.\right].$$

Proof: The left–hand side in (III.3.1)$_\pm$ is equal to

$$\varphi(z,s) - \varphi(z',s') - \mathscr{R}e\,[A\cdot(z-z')] + \beta(s-s'+\mathscr{I}m\,[A\cdot(z-z')])$$

which, if (III.2.1')$_+$ holds, is

$$\geq \varphi(z,s) - \varphi(z,s'-\mathscr{I}m\,[A\cdot(z-z')]) -$$

$$(s-s'+\mathscr{I}m\,[A\cdot(z-z')])\int_0^1 \varphi_s(z,s+\lambda(s'-s-\mathscr{I}m\,[A\cdot(z-z')]))\mathrm{d}\lambda \equiv 0\ .$$

If $(III.2.1')_-$ holds the sign $\geq$ must be replaced by $\leq$. ⬤

COROLLARY III.3.1.– *Suppose the map A satisfies conditions* $(III.2.1')_-$ *(resp.,* $(III.2.1')_+$*) & (III.2.2) and set* $B(z,s,z') = A(z',z,s)$*. Then*

$(III.3.2)_+$
$$\mathscr{I}m\ \left[[1+\imath\beta(z',s',z,s)][s+\imath\varphi(z,s)-s'-\imath\varphi(z',s')-\imath B(z,s,z')\cdot(z-z')]\right]\ \geq\ 0$$

$\left[resp.,\right.$
$(III.3.2)_-$
$$\mathscr{I}m\ \left[[1+\imath\beta(z',s',z,s)][s+\imath\varphi(z,s)-s'-\imath\varphi(z',s')-\imath B(z,s,z')\cdot(z-z')]\right]\ \leq\ 0\left.\right]\ ;$$

$(III.3.3)$ $$\mathrm{rank}\ (\partial B_i/\partial\overline{z'_j})_{1\leq i,j\leq n}\ \leq\ n-\nu\ .$$

We are now in a position to apply Th. II.7.4. (we are assuming throughout that diam $\mathcal{O}$ is suitably small). We select, when $\sigma > 0$,

$$\mathbf{a}\left.\right|_{t=0}\ =\ \sigma A\ \ if\ (III.2.1')_+\ holds,\ \ \mathbf{a}\left.\right|_{t=0}\ =\ \sigma B\ \ if\ (III.2.1')_-\ holds\ .$$

In the former case we select $\mu = 1 + \imath\beta(z,s,z',s')$; in the latter case, $\mu = 1 + \imath\beta(z',s',z,s)$. When $\sigma < 0$, we take

$$\mathbf{a}\left.\right|_{t=0}\ =\ \sigma B\ \ if\ (III.2.1')_+\ holds,\ \ \mathbf{a}\left.\right|_{t=0}\ =\ \sigma A\ \ if\ (III.2.1')_-\ holds,$$

with $\mu = 1 + \imath\beta(z',s',z,s)$ or $\mu = 1 + \imath\beta(z,s,z',s')$ respectively.

Clearly, the definition of $\mathbf{a}$ satisfies the requirements in Sect. II.5 and that of μ satisfies condition (II.6.5), provided, of course, diam($\mathcal{D} \times \mathcal{B}$) is small enough (notice that $|\beta| \to 0$ as diam($\mathcal{D} \times \mathcal{B}$) $\to 0$, by virtue of the fact that $\varphi_s|_0 = 0$).

By Prop. III.3.1 & Cor. III.3.1 the inequality (II.6.6) is valid in all cases. Property (III.2.2) implies the validity of (II.2.10) if $q > n - \nu$; property (III.3.3), implies that of (II.2.11) if $q < \nu$. We may state:

THEOREM III.3.1.– *If Σ admits supporting manifolds from above, of holomorphic type one, then we have, in $\mathring{O}$,*

(III.3.4) $I_{\theta}^{+}f = K_{O}^{+}df - K_{\partial\Omega\times\theta}^{+}f , \quad \forall \ f \ \in \ C^{\infty}(\mathscr{E}\angle\mathring{O};\Lambda^{n+1,0}).$

If Σ admits supporting manifolds from below, of holomorphic type one, then we have, in $\mathring{O}$,

(III.3.5) $I_{\theta}^{+}f = (-1)^{n-1}d(K_{O}^{+}f) , \quad \forall \ f \ \in \ C^{\infty}(\mathscr{E}\angle\mathring{O};\Lambda^{n+1,n}).$

A similar statement, with superscripts − , is valid if we exchange "above" and "below".

COROLLARY III.3.2.– *Suppose Σ admits supporting manifolds both from above and from below, of holomorphic type one. Then we have, in $\mathring{O}$,*

(III.3.6) $f = K_{O}df - K_{\partial\Omega\times\theta}f , \forall \ f \ \in \ C^{\infty}(\mathscr{E}\angle\mathring{O};\Lambda^{n+1,0});$

(III.3.7) $f = (-1)^{n-1}d(K_{O}f) , \forall \ f \ \in \ C^{\infty}(\mathscr{E}\angle\mathring{O};\Lambda^{n+1,n}).$

We have used the notation $K = K^{+} + K^{-}$.

We propose now to apply Theorems II.7.5 & II.7.6 under hypotheses similar to those in Th. III.3.1. Therefore we limit our attention to the cases

(III.3.8) $1 \leq q \leq n - 1 .$

We begin by reasoning under the following hypothesis:

(III.3.9)   *the hypersurface Σ admits supporting manifolds from below, of holomorphic type ν_+ , and from above, of holomorphic type ν_- .*

Note that if $q > n - \nu_+$ property (II.2.10), and therefore also (II.5.10), is satisfied when we take $a = \sigma A$ with A as in Def. III.2.1. This suggest that we take a inde– pendent of t , which has the added advantage of ensuring trivially the validity of (II.5.11). As for condition (II.6.11) it is an immediate consequence of (III.3.1)$_+$; and there is no need to take $\hat{O}_0$ smaller than $\hat{O}$. We may state

THEOREM III.3.2.– *If* (III.3.9) *holds and if* $q > n - \nu_+$ *, then the homotopy formula* (II.7.7) *is valid in the open set* $\hat{O}$ *, for all differential forms* $f \in C^\infty(\mathscr{C}\!\angle\hat{O};\Lambda^{n+1,q})$.

The formula analogous to (II.7.7) with superscripts – is valid when $q > n - \nu_-$.

     We must now look at the values $q \leq n - \nu_+$ or $q \leq n - \nu_-$. This is where the concept of strictly supporting manifolds turns out to be handy.

PROPOSITION III.3.2. – *Call* $\varpi(z,z',s')$ *the left–hand side in* (III.2.6')$_+$ (*or in* (III.2.6')$_-$) *and set*

$$\beta_+(z,s,z',s') = s - s' + \mathcal{Jm}\,[A(z,z',s')\cdot(z-z')] -$$

$$\int_0^1 \varphi_s(z,s+\lambda(s'-s-\mathcal{Jm}\,[A(z,z',s')\cdot(z-z')]))\mathrm{d}\lambda \;;$$

$$\beta_-(z,s,z',s') = - (s-s'+\mathcal{Jm}\,[A(z,z',s')\cdot(z-z')]) -$$

$$\int_0^1 \varphi_s(z,s+\lambda(s'-s-\mathcal{Jm}\,[A(z,z',s')\cdot(z-z')]))\mathrm{d}\lambda \;.$$

Then, for all $(z,s,z',s') \in (\mathscr{D} \times \mathscr{B})\times(\mathscr{D} \times \mathscr{B})$,

(III.3.10)$_+$

$$\mathcal{I}m\left[[1+\imath\beta_+(z,s,z',s')][s+\imath\varphi(z,s)-s'-\imath\varphi(z',s')-\imath A(z,z',s')\cdot(z-z')]\right] =$$

$$- \varpi(z,z',s') + \left|s-s'+\mathcal{I}m\,[A(z,z',s')\cdot(z-z')]\right|^2 ;$$

(III.3.10)$_-$

$$- \mathcal{I}m\left[[1+\imath\beta_-(z,s,z',s')][s+\imath\varphi(z,s)-s'-\imath\varphi(z',s')-\imath A(z,z',s')\cdot(z-z')]\right] =$$

$$\varpi(z,z',s') + \left|s-s'+\mathcal{I}m\,[A(z,z',s')\cdot(z-z')]\right|^2 .$$

$\mathcal{P}roof$: The left–hand side in (III.3.10)$_+$ is equal to

$$\varphi(z,s) - \varphi(z',s') - \mathcal{R}e\,[A\cdot(z-z')] + \beta_+[s-s'+\mathcal{I}m\,A\cdot(z-z')] =$$

$$\varphi(z,s) - \varpi(z,z',s') - \varphi(z,s'-\mathcal{I}m\,[A\cdot(z-z')]) + \left|s-s'+\mathcal{I}m\,[A(z,z',s')\cdot(z-z')]\right|^2 -$$

$$(s-s'+\mathcal{I}m\,[A\cdot(z-z')])\int_0^1 \varphi_s(z,s+\lambda(s'-s-\mathcal{I}m\,[A\cdot(z-z')]))\mathrm{d}\lambda =$$

$$- \varpi(z,z',s') + \left|s-s'+\mathcal{I}m\,[A(z,z',s')\cdot(z-z')]\right|^2 .$$

Likewise for (III.3.10)$_-$. ●

COROLLARY III.3.3.– *Set* $B(z,s,z') = A(z',z,s)$. *Then we have, in* $(\mathcal{D}\times\mathcal{B})\times(\mathcal{D}\times\mathcal{B})$,

(III.3.11)$_+$

$$\mathcal{I}m\left[[1+\imath\beta_-(z',s',z,s)][s+\imath\varphi(z,s)-s'-\imath\varphi(z',s')-\imath B(z,s,z')\cdot(z-z')]\right] =$$

$$- \varpi(z',z,s) + \left|s-s'+\mathcal{I}m\,[B(z,s,z')\cdot(z-z')]\right|^2 ;$$

(III.3.11)$_-$

$$- \mathcal{I}m \left[[1+\imath\beta_+(z',s',z,s)][s+\imath\varphi(z,s)-s'-\imath\varphi(z',s')-\imath B(z,s,z')\cdot(z-z')] \right] =$$

$$\varpi(z',z,s) + \left| s-s'+\mathcal{I}m \left[B(z,s,z')\cdot(z-z') \right] \right|^2 .$$

We let the maps A & B be defined as in Cor. III.3.3. We assume that A satis- fies (III.2.2), and therefore (Prop. III.3.1) B satisfies (III.3.3). We derive that if we take $a\big|_{t=0} = \sigma B$ and if A satisfies (III.2.1')$_-$ then (II.6.6) (at $t = 0$) holds for $\sigma > 0$, whereas if A satisfies (III.2.1')$_+$ then (II.6.6) holds for $\sigma < 0$.

Let us now assume that (III.2.6')$_+$ (resp. (III.2.6')$_-$) holds, i. e., $\varpi(z,z',s') > 0$ (resp. $\varpi(z',z,s) > 0$) unless $z = z'$.

Let us provisionally restrict the variation of (z,s) to $\hat{\mathcal{O}}$ and that of (z',s') to the boundary $\partial\hat{\mathcal{O}}$. When $z = 0$, $s = 0$, $\varpi(0,z',s') \neq 0$ unless $z' = 0$ which demands $s' \in \partial\Theta$ and therefore $|s'|^2 > 0$. Thus the right–hand side in (III.3.11)$_+$ is bounded away from zero, from below. By compactness and continuity this remains true if we allow (z,s) to vary in some open neighborhood $\hat{\mathcal{O}}_0 \subset \hat{\mathcal{O}}$ of the origin. We get, for some positive constants C_0 , c_1 and any $t \in [0,1]$,

$$\mathcal{I}m \left[[1+\imath\beta_-(z',s',z,s)][s+\imath\varphi(z,s)-s'-\imath\varphi(z',s')-\imath B((1-t)z,(1-t)s,z')\cdot(z-z')] \right] \geq$$

$$\mathcal{I}m \left[[1+\imath\beta_-(z',s',z,s)][s+\imath\varphi(z,s)-s'-\imath\varphi(z',s')-\imath B(z,s,z')\cdot(z-z')] \right] -$$

$$C_0(|z|^2+|s|^2)^{\frac{1}{2}} \geq c_1 - C_0(|z|^2+|s|^2)^{\frac{1}{2}}.$$

A similar remark applies to the right–hand sides in (III.3.11)$_-$. We further contract $\hat{\mathcal{O}}_0$ about the origin, to ensure diam $\hat{\mathcal{O}}_0 < c_1/C_0$, and define

$$a = \sigma B((1-t)z,(1-t)s,z') .$$

It follows that, when A satisfies (III.2.6')$_-$ then (II.6.11) holds for $\sigma > 0$, whereas if A satisfies (III.2.6')$_+$ then (II.6.11) holds for $\sigma < 0$.

In view of the choice of a condition (II.2.10) is trivially satisfied when $t = 1$.

Lastly we take a look at conditions (II.5.10) & (II.5.11). The information at our disposal is embodied in property (III.3.3). The latter entails

$$(\text{III.3.12}) \qquad \bigwedge_{j \in J} L' a_j \equiv 0 \quad \text{as soon as } |J| > n - \nu ,$$

which entails (II.5.10) provided $q < \nu$, and (II.5.11) provided $q < \nu - 1$.

We are now in a position to apply Theorems II.7.5 & II.7.6:

THEOREM III.3.3.– *Assume* $1 \leq q \leq n$ *and that*

(III.3.13)   *the hypersurface* Σ *admits strictly supporting manifolds from below, of holomorphic type* ν_+ , *and from above, of holomorphic type* ν_- .

If $q < \nu_- - 1$ *the homotopy formula* (II.7.7) *is valid in the open set* $\hat{O}_0$, *for all forms* $f \in C^\infty(\mathscr{E} \angle \hat{O}; \Lambda^{n+1,q})$. *If* $q < \nu_+ - 1$ *the same is true with superscripts* $-$.

If $q < \nu_-$ *the right–inverse formula* (II.7.8) *is valid in the open set* $\hat{O}_0$, *for all closed forms* $f \in C^\infty(\mathscr{E} \angle \hat{O}; \Lambda^{n+1,q})$. *If* $q < \nu_+$ *the same is true with superscripts–*.

We may of course combine Theorems III.3.2 & III.3.3:

COROLLARY III.3.4.– *Same hypotheses as in Th. III.3.3.*

If $q \notin [\nu_- - 1, n - \nu_+]$ *the homotopy formula* (II.7.7) *is valid in the open set* $\hat{O}_0$, *for all forms* $f \in C^\infty(\mathscr{E} \angle \hat{O}; \Lambda^{n+1,q})$. *If* $q \notin [\nu_+ - 1, n - \nu_-]$ *the same is true with superscripts* $-$.

If $q \notin [\nu_-, n - \nu_+]$ *the right–inverse formula* (II.7.8) *is valid in the open set* $\hat{O}_0$, *for all closed forms* $f \in C^{\infty}(\mathscr{C}\angle\hat{\mathcal{O}};\Lambda^{n+1,q})$. *If* $q \notin [\nu_+, n-\nu_-]$ *the same is true with superscripts* $-$.

COROLLARY III.3.5.– *Same hypotheses as in Th. III.3.3.*

If $q \notin [\nu_- -1, n-\nu_+] \cup [\nu_+ -1, n-\nu_-]$ *then, for all forms* $f \in C^{\infty}(\mathscr{C}\angle\hat{\mathcal{O}};\Lambda^{n+1,q})$ *we have, in the open set* $\hat{\mathcal{O}}_0$,

$$(\text{III.3.14}) \qquad (-1)^q f = \mathrm{d}[(\mathbf{K}_{\hat{\mathcal{O}}} - \mathbf{B}_{\partial\Omega\times\theta})f] + (\mathbf{K}_{\hat{\mathcal{O}}} - \mathbf{B}_{\partial\Omega\times\theta})\mathrm{d}f .$$

If $q \notin [\nu_-, n-\nu_+] \cup [\nu_+, n-\nu_-]$ *then, for all closed forms* $f \in C^{\infty}(\mathscr{C}\angle\hat{\mathcal{O}};\Lambda^{n+1,q})$ *we have, in the open set* $\hat{\mathcal{O}}_0$,

$$(\text{III.3.15}) \qquad f = \mathrm{d}[(-1)^q(\mathbf{K}_{\hat{\mathcal{O}}} - \mathbf{B}_{\partial\Omega\times\theta})f] .$$

As usual we have written $\mathbf{K}_{\hat{\mathcal{O}}} = \mathbf{K}_{\hat{\mathcal{O}}}^+ + \mathbf{K}_{\hat{\mathcal{O}}}^-$, $\mathbf{B}_{\partial\Omega\times\theta} = \mathbf{B}_{\partial\Omega\times\theta}^+ + \mathbf{B}_{\partial\Omega\times\theta}^-$.

REMARK III.3.1.– Prop. III.2.2 shows that property (III.3.13), and *a fortiori* property (III.2.6), holds when ν_+ (resp., ν_-) is the number of strictly positive (resp., negative) eigenvalues of the $n \times n$ matrix $\varphi_{z\bar{z}}(0,0)$. The preceding statements are thus consistent with the classical results (see [A–H]). ☻

REMARK III.3.2.– Consider the case $\nu_+ = n$ (thus $\nu_- = 0$). Cor. III.3.5 provides a right–inverse formula, but no local homotopy formula, when $q = n-1$. In particular this applies in the strongly pseudoconvex case when $n = 2$ (Σ is a real hypersurface in $\mathbf{C}^3$ and all the eigenvalues of the Levi form of Σ at the origin are different from zero and have the same sign) and $q = 1$. This is consistent with the example in [Na–R]. ☻

III.4.   LOCAL HOMOTOPY FORMULAS IN HIGHER CODIMENSION

We now consider the case $d \geq 2$. We denote by Σ the generic submanifold of $\mathbb{C}^{n+d}$ defined by the equations (II.1.35): Σ is the image of $\mathscr{D} \times \mathscr{B}$ under the map $(z,s) \rightarrow (z,w)$ with w given by (II.1.2); we assume that (II.1.3) holds. Throughout the present section we reason under hypothesis (II.4.1).

We denote by T^0 the characteristic set in the CR structure on Σ , and we identify its fibre at the origin to $\mathbb{R}^d$ via the isomorphism

$$\mathbb{R}^d \ni \sigma = (\sigma_1 ,...,\sigma_d) \rightarrow \sum_{k=1}^{d} \sigma_k \mathrm{d}s_k$$

(see end of Sect. II.1). We recall [see (II.1.34)] that the Levi matrix at the characteristic point $(0,\sigma)$ can be identified to the complex Hessian

$$\partial\overline{\partial}(\sigma \cdot \varphi)\big|_0 = \sum_{k=1}^{d} \sigma_k \left[\partial^2\varphi_k / \partial z_i \partial\overline{z}_j\right]_{1 \leq i,j \leq n} .$$

The Levi matrix at $(0,\sigma)$ depends linearly on σ .

The **forthcoming analysis** will be based on the following hypothesis:

(III.4.1)   *There are exactly ν strictly positive eigenvalues of the Levi matrix at every point $(0,\sigma) \in T^0$, $\sigma \neq 0$.*

The following example shows that hypothesis (III.4.1) is not void:

EXAMPLE III.4.1.– Suppose $d = 2$ and

(III.4.2) $\varphi_1 = 2\mathscr{R}e(z_1\overline{z}_2) , \quad \varphi_2 = 2\mathscr{I}m(z_1\overline{z}_2) .$

Write $\sigma = |\sigma|e^{i\theta}$; then $\sigma \cdot \varphi = |\sigma| \mathcal{R}e \,(z_1 \bar{z}_2 e^{-i\theta})$ and the Levi matrix at the origin can be identified to the 2×2 matrix

$$|\sigma| \begin{bmatrix} 0 & e^{-i\theta} \\ e^{i\theta} & 0 \end{bmatrix} \,,$$

whose eigenvalues are $|\sigma|$ & $-|\sigma|$. ❧

PROPOSITION III.4.1.– *If* (III.4.1) *holds there are exactly ν strictly negative eigenvalues of the Levi matrix at every point $(0,\sigma) \in T^0$, $\sigma \neq 0$, and exactly $n - 2\nu$ eigenvalues which are equal to zero.*

Proof : If $\lambda(\sigma)$ is an eigenvalue of the Levi matrix at $(0,\sigma)$ then $-\lambda(\sigma)$ is an eigenva– lue of the Levi matrix at $(0,-\sigma)$. ❧

In particular, if (III.4.1) holds then necessarily $2\nu \leq n$. The next statement is well–known; we include a proof to make the on–going argument self–contained.

PROPOSITION III.4.2.– *There are n real–valued, continuous functions $\lambda_1 ,..., \lambda_n$ in $\mathbb{R}^d$ which represent, at each point $\sigma \in \mathbb{R}^d$, the eigenvalues of the matrix $\partial\bar{\partial}(\sigma \cdot \varphi)\big|_0$. Furthermore they can be taken to be positive–homogeneous of degree one and labeled in such a manner that $\lambda_1 \leq ... \leq \lambda_n$.*

Proof : It will suffice to define the functions λ_i on the unit sphere S^{d-1}. At each point σ they are the roots of the characteristic polynomial $P(\sigma;\lambda)$ of the matrix $\partial\bar{\partial}(\sigma \cdot \varphi)\big|_0$. We can as well assume that the assertion to be proved concerns the roots of an arbitrary monic polynomial of degree n ,

$$P(\sigma;\lambda) = \lambda^n + a_1(\sigma)\lambda^{n-1} +...+ a_n(\sigma) \,,$$

whose **coefficients** $a_j(\sigma)$ are continuous functions of $\sigma \in S^{d-1}$ and whose roots are real. The assertion is trivial when $n = 1$. Suppose $n \geq 2$ and let $\Lambda(\sigma)$ denote the largest value of a root of $P(\sigma;\lambda)$. We contend that Λ is a continuous function in S^{d-1}. Indeed, suppose it were not so: there would be a sequence of points σ^ι ($\iota = 1,2,...$) converging to a point σ_0 such that $\Lambda(\sigma^\iota)$ converge to a root $\lambda_* < \Lambda(\sigma_0)$ of $P(\sigma_0;\lambda)$. If the number $\varepsilon > 0$ is sufficiently small and if $|\sigma_0-\sigma| < \varepsilon$ then there is a root $\lambda(\sigma)$ of $P(\sigma;\lambda)$ such that

$$|\lambda(\sigma) - \Lambda(\sigma_0)| < \tfrac{1}{2}(\Lambda(\sigma_0) - \lambda_*) .$$

On the other hand, if ι is large enough, we have

$$|\Lambda(\sigma^\iota) - \lambda_*| < \tfrac{1}{2}(\Lambda(\sigma_0) - \lambda_*) ,$$

and also $|\sigma^\iota-\sigma_0| < \varepsilon$, whence

$$\lambda(\sigma^\iota) - \Lambda(\sigma^\iota) \geq \Lambda(\sigma_0) - \lambda_* - |\Lambda(\sigma^\iota) - \lambda_*| - |\Lambda(\sigma_0) - \lambda(\sigma^\iota)| > 0 ,$$

which contradicts the definition of Λ .

Define a polynomial

$$Q(\sigma;\lambda) = \lambda^{n-1} + b_1(\sigma)\lambda^{n-2} +...+ b_{n-1}(\sigma)$$

by the relation

$$P(\sigma;\lambda) = (\lambda - \Lambda(\sigma))Q(\sigma;\lambda) ,$$

which means that

$$b_i - \Lambda b_{i-1} = a_i \quad , \; i = 1,...,n-1 ,$$

with the understanding that $b_0 = 1$. By increasing induction on i we conclude that each coefficient b_i is a continuous function of σ in S^{d-1}. Every root of Q is a root of P ; therefore, induction on deg P applies. ●

The conjunction of Propositions III.4.1 & III.4.2 allows us to state:

PROPOSITION III.4.3.– *Suppose* (III.4.1) *holds, and let* $\lambda_1 \leq ... \leq \lambda_n$ *be the eigenvalues of* $\partial\bar{\partial}(\sigma\cdot\varphi)\big|_0$. *There is a number* $\eta > 0$ *such that, in* S^{d-1},

$$- \eta^{-1} \leq \lambda_1 \leq...\leq \lambda_\nu \leq - \eta \ , \ \lambda_{\nu+1} =...= \lambda_{n-\nu-1} = 0 \ ,$$

(III.4.3)

$$\eta \leq \lambda_{n-\nu} \leq...\leq \lambda_n \leq \eta^{-1} \ .$$

We select a smooth closed curve $\mathfrak{c}$ in the half–plane $\mathcal{R}e\, \lambda > 0$ which winds once around the closed interval $[\eta,\eta^{-1}] \subset \,]0,+\infty[$ and we set, for any $\sigma \in S^{d-1}$,

(III.4.4) $$A^+(\sigma) = (2\imath\pi)^{-1}\oint_{\mathfrak{c}} [\lambda I_n - \partial\bar{\partial}(\sigma\cdot\varphi)]^{-1}\sqrt{\lambda} \ d\lambda \ ,$$

where $\sqrt{\lambda}$ is the branch of the square–root function that is positive on the positive half of the real axis and I_n is the identity $n{\times}n$ matrix. For $\sigma \in \mathbb{R}^d\backslash\{0\}$ we set

$$A^+(\sigma) = |\sigma|^{\frac{1}{2}}A^+(\sigma/|\sigma|) \ ;$$

A^+ extends as a continuous function of $\sigma \in \mathbb{R}^d$, positive–homogeneous of degree $\frac{1}{2}$. Likewise, by substituting $-\mathfrak{c}$ for $\mathfrak{c}$ in (III.4.4) and taking now $\sqrt{\lambda}$ to mean the branch that is positive on the negative half of the real axis we define $A^-(\sigma)$. Both A^+ & A^- are nonnegative self–adjoint $n{\times}n$ matrices. We have

(III.4.5) $$\partial\bar{\partial}(\sigma\cdot\varphi)\big|_0 = (A^+)^2 - (A^-)^2 \ .$$

Next we simplify the expressions of the functions φ_k by adapting the reasoning developed in Sect. III.2 in the case $d = 1$. Let the functions w_k be given by (II.1.2). We select coefficients $a_{hi,\ell}$, $b_{ij,\ell}$ & $c_{jk,\ell}$ $(1 \leq h$, $i \leq n$, $1 \leq j$, k, $\ell \leq d)$ such that, if we set

$$\tilde{w}_\ell = w_\ell - \sum_{h,i=1}^{n} a_{hi,\ell} z_h z_i - \sum_{i=1}^{n} \sum_{j=1}^{d} b_{ij,\ell} z_i w_j - \sum_{j,k=1}^{d} c_{jk,\ell} w_j w_k ,$$

then

$$\tilde{s}_\ell = \mathcal{R}e\, \tilde{w}_\ell = s_\ell + O(|z|^2 + |s|^2) ,$$

$$\tilde{\varphi}_k = \mathcal{I}m\, \tilde{w}_k = \varrho_k(z) + O(|z|^3 + |s|^3) ,$$

and $\tilde{\varphi}_k = \mathcal{I}m\, \tilde{w}_k$ is congruent, modulo terms that vanish at least to third order at the origin, to a hermitian quadratic form ϱ_k in $\mathbf{C}^n$. In other words, after deleting the tildas, we may assume that

$$(III.4.6) \qquad \varphi_k = \varrho_k(z) + O(|z|^3 + |s|^3) , \quad k = 1,...,n .$$

We have

$$(III.4.7) \qquad \sum_{k=1}^{d} \sigma_k \varrho_k(z) = \bar{z} \cdot \left[\partial \bar{\partial}(\sigma \cdot \varphi)\big|_0 z \right] = |A^+(\sigma)z|^2 - |A^-(\sigma)z|^2 .$$

In the sequel we write $\dot{\sigma} = \sigma/|\sigma| \in S^{d-1}$ $(\sigma \neq 0)$. We define

$$\mu_+ = [I_d + \iota\, {}^t\varphi_s(z',s')]^{-1}[I_d + \iota(s-s')\circ\dot{\sigma}] ,$$

$$(III.4.8)$$

$$\mu_- = [I_d + \iota\, {}^t\varphi_s(z,s)]^{-1}[I_d + \iota(s-s')\circ\dot{\sigma}] ,$$

where I_d is the $d \times d$ identity matrix and where we have denoted by $(s{-}s') \circ \dot{\sigma}$ the matrix which transforms a d–vector $\mathbf{u}$ into $(\dot{\sigma} \cdot \mathbf{u})(s{-}s')$. Note that the transpose of $(s{-}s') \circ \dot{\sigma}$ is nothing else but $\dot{\sigma} \circ (s{-}s')$. The $d \times d$ matrices μ_+ and μ_- define linear endomorphisms of $\mathbb{C}^d$ which conform to the requirements in (II.6.5).

LEMMA III.4.1.– *Given any* $\varepsilon > 0$, *provided* diam $(\mathscr{D} \times \mathscr{B})$ *is small enough, we have, in* $\mathscr{D} \times \mathscr{B} \times S^{d-1}$,

$$(III.4.9) \qquad \mathscr{I}m\left[\mu_+ \dot{\sigma} \cdot [s + \imath \varphi(z,s) - s' - \imath \varphi(z',s')]\right] \geq$$

$$\mathscr{R}e\left[\mu_+ \dot{\sigma} \cdot [\varphi(z,s) - \varphi(z',s)]\right] + (1{-}\varepsilon)|s{-}s'|^2 ;$$

$$(III.4.10) \qquad \mathscr{I}m\left[\mu_- \dot{\sigma} \cdot [s + \imath \varphi(z,s) - s' - \imath \varphi(z',s')]\right] \geq$$

$$\mathscr{R}e\left[\mu_- \dot{\sigma} \cdot [\varphi(z,s') - \varphi(z',s')]\right] + (1{-}\varepsilon)|s{-}s'|^2 .$$

Proof : Thanks to the hypothesis that $\|d\varphi\|$ is as small as desired provided the open set $\mathscr{D} \times \mathscr{B}$ is sufficiently small, we can obtain

$$\|I_d + \imath(s{-}s') \circ \dot{\sigma}\| \left| [I_d + \imath \varphi_s(z',s')]^{-1}[s + \imath \varphi(z,s) - s' - \imath \varphi(z',s')] - \right.$$

$$\left. (s{-}s') - \imath[I_d + \imath \varphi_s(z',s')]^{-1}[\varphi(z,s) - \varphi(z',s)] \right| \leq \varepsilon |s{-}s'|^2.$$

On the other hand,

$$\mathscr{I}m\left[[I_d + \imath \dot{\sigma} \circ (s{-}s')]\{(s{-}s') + \imath[1 + \imath \varphi_s(z',s')]^{-1}[\varphi(z,s) - \varphi(z',s)]\}\right] =$$

$$|s{-}s'|^2 \dot{\sigma} + \mathscr{R}e\left[{}^t\mu_+[\varphi(z,s) - \varphi(z',s)]\right] ,$$

whence (III.4.9). The inequality (III.4.10) is derived in similar fashion. ●

LEMMA III.4.2.- *Given any* $\varepsilon > 0$, *provided* diam $(\mathscr{D} \times \mathscr{B})$ *is small enough, there exist* C^{∞} *maps* a_{+} , $a_{-} : (\mathscr{D} \times \mathscr{B}) \times (\mathscr{D} \times \mathscr{B}) \times S^{d-1} \times [0,1] \to \mathbb{C}^{n}$, *which extend as holomorphic functions of* $\dot{\sigma}$ *in some open neighborhood of* S^{d-1} *in* $\mathbb{C}^{d}$ *and are such that, for some constants* K , $c_{0} > 0$ *and all* $(z,s,z',s',\sigma,t) \in (\mathscr{D} \times \mathscr{B}) \times (\mathscr{D} \times \mathscr{B}) \times S^{d-1} \times [0,1]$,

(III.4.11)
$$\mathscr{R}e\left[\mu_{+}\dot{\sigma}\cdot[\varphi(z,s)-\varphi(z',s)] - a_{+}(z,s,z',s',\mu_{+}\dot{\sigma},t)(z-z')\right] \geq$$

$$c_{0}|z-z'|^{2} - \varepsilon|s-s'|^{2} - Kt|z| ;$$

(III.4.12)
$$\mathscr{R}e\left[\mu_{-}\dot{\sigma}\cdot[\varphi(z,s')-\varphi(z',s')] - a_{-}(z,s,z',s',\mu_{-}\dot{\sigma},t)(z-z')\right] \geq$$

$$c_{0}|z-z'|^{2} - \varepsilon|s-s'|^{2} - Kt(|z|^{2}+|s|^{2})^{\frac{1}{2}}.$$

Furthermore we can achieve that a_{+} (resp., a_{-}) *be independent of* s (resp., s') *and*

(III.4.13)
$$\text{rank } (\partial a_{+}/\partial \bar{z}) = \text{rank } (\partial a_{-}/\partial \bar{z}') = n - \nu ;$$

(III.4.14)
$$a_{+}\big|_{t=1} \ \& \ a_{-}\big|_{t=1} \ \textit{are independent of} \ (z,s) .$$

Proof: We shall determine a_{+} ; the determination of a_{-} is analogous. We are going to exploit (III.4.6), which we rewrite as follows:

(III.4.15)
$$\varphi(z,s) = Q(z) + \psi(z,s) ,$$

with $|\psi| \leq const.(|z|^{3}+|s|^{3})$. We begin by observing that, if diam $(\mathscr{D} \times \mathscr{B})$ is small enough, then

(III.4.16)
$$\left|\psi(z,s) - \psi(z',s) - 2\mathscr{R}e\left[\psi_{z}(z',s')(z-z')\right]\right| \leq$$

$$\varepsilon^2(|z-z'|^2 + |s-s'|^2) \ .$$

Set

$$\tilde{\mu}_+ = \mu_+[I_d + \imath(s-s')\circ\dot{\sigma}]^{-1}, \ \gamma_+(z',s') = [I_d - \imath\varphi_s(z',s')]^{-1}\psi_z(z',s') \ .$$

We contend that it suffices to choose a_+ independent of s and such that

(III.4.17)
$$a_+\big|_{z=z'=0,s'=0} \equiv 0 \ ,$$

(III.4.18)
$$\mathcal{R}e\left[\tilde{\mu}_+\dot{\sigma}\cdot[\mathcal{Q}(z)-\mathcal{Q}(z')+2\gamma_+(z',s')(z-z')] - a_+(z,z',s',\tilde{\mu}_+\dot{\sigma},t)\cdot(z-z')\right] \geq$$

$$c_0|z-z'|^2 - Kt|z| \ .$$

[Throughout the argument ε will be very small compared to c_0 .]

Note first that

$$\mathcal{R}e\left[\tilde{\mu}_+\dot{\sigma}\cdot\gamma_+(z',s')(z-z')\right] = [I_d + {}^t\varphi_s^2(z',s')]^{-1}\dot{\sigma}\cdot\mathcal{R}e\left[\psi_z(z',s')(z-z')\right] \ ,$$

which, **thanks** to (III.4.16) and to the fact that $\mathcal{R}e\,\tilde{\mu}_+ = [I_d + {}^t\varphi_s^2(z',s')]^{-1}$, shows that

$$\left|\mathcal{R}e\left[[2\tilde{\mu}_+\dot{\sigma}\cdot\gamma_+(z',s')(z-z')] - \tilde{\mu}_+\dot{\sigma}\cdot[\psi(z,s)-\psi(z',s)]\right]\right| \leq$$

$$2\varepsilon^2(|z-z'|^2+|s-s'|^2)$$

provided diam $(\mathcal{D}\times\mathcal{B})$ is small enough. Choosing diam $(\mathcal{D}\times\mathcal{B})$ small enough allows us also to achieve

$$|(\mu_+ - \tilde{\mu}_+)\dot{\sigma} \cdot [\varphi(z,s) - \varphi(z',s)]| \leq \varepsilon^2(|z-z'|^2 + |s-s'|^2) ,$$

$$|[a_+(z,z',s',\mu_+\dot{\sigma},t) - a_+(z,z',s',\tilde{\mu}_+\dot{\sigma},t)](z-z')| \leq \varepsilon^2(|z-z'|^2 + |s-s'|^2) ,$$

the latter thanks to (III.4.17). Putting all these inequalities together, selecting $\varepsilon > 0$ suitably small and redefining c_0 yields (III.4.11).

The next step is to observe that it suffices to determine a map $\mathcal{d}_+$ similar to a_+ but submitted to the requirement that

(III.4.19)
$$\mathcal{R}e\left[\tilde{\mu}_+\dot{\sigma} \cdot [Q(z) - Q(z')] - \mathcal{d}_+(z,z',s',\tilde{\mu}_+\dot{\sigma},t) \cdot (z-z')\right] \geq$$

$$c_0|z-z'|^2 - Kt|z| .$$

For then the map

$$a_+ = \mathcal{d}_+ + 2\tilde{\mu}_+\dot{\sigma} \cdot \gamma_+(z',s')$$

will satisfy all our *desiderata*.

Observe then that

$$\mathcal{R}e\left[\tilde{\mu}_+\dot{\sigma} \cdot [Q(z) - Q(z')]\right] = \mathcal{R}e\left[\sigma' \cdot [Q(z) - Q(z')]\right] ,$$

where $\sigma' = [I_d + {}^t\varphi_s(z',s')^2]^{-1}\dot{\sigma}$ ($\in \mathbb{R}^d$). We continue to take diam ($\mathscr{D} \times \mathscr{B}$) as small as deemed necessary. Suppose we determine a map $\mathcal{b}$ such that

(III.4.20)
$$\mathcal{R}e\left[\sigma' \cdot [Q(z) - Q(z')] - \mathcal{b}(z,z',\sigma',t) \cdot (z-z')\right] \geq$$

$$c_0|z-z'|^2 - Kt|z| .$$

Then the map

$$\mathcal{A}_+(z,z',s',\sigma,t) \; = \; \mathcal{A}(z,z',[I_d - i\,{}^t\varphi_s(z',s')]^{-1}\sigma,t)$$

will satisfy (III.4.19).

The remainder of the proof is devoted to the determination of the map $\mathcal{A}$. First note that, by (III.4.4) and the analogous formula with $-c$ in the place of c, A^+ & A^- extend as holomorphic functions of σ (valued in the space of complex $n \times n$ matrices) in an open neighborhood $\mathcal{U}$ of S^{d-1} in $\mathbf{C}^d$. Let c_0 be a circle in the complex plane, with center 0 and radius $< \eta/2$. The following $n \times n$ matrix depends holomorphically on $\sigma \in \mathcal{U}$ (possibly contracted about S^{d-1}):

(III.4.21) $$P_0(\sigma) \; = \; (2i\pi)^{-1} \oint_{c_0} [\lambda I_n - \partial\bar{\partial}(\sigma \cdot \varphi)]^{-1} d\lambda \; .$$

When $\sigma \in \mathcal{U} \cap \mathbf{R}^d$ $P_0(\sigma)$ is the orthogonal projector on Ker $\partial\bar{\partial}(\sigma \cdot \varphi)$. We then introduce, for $\sigma \in \mathcal{U}$,

(III.4.22)

$$\mathcal{A}(z,z',\sigma) \; = \; 2\,{}^t A^+(\sigma){}^t A^+(\sigma)\bar{z}' \; - \; 2\,{}^t A^-(\sigma){}^t A^-(\sigma)\bar{z} \; - \; {}^t P_0(\sigma)(\bar{z}-\bar{z}') \; .$$

We recall that, for $\sigma \in \mathcal{U} \cap \mathbf{R}^d$, $A^\pm(\sigma)$ is self-adjoint; we have, then,

$$\sigma \cdot [Q(z) - Q(z')] \; - \; \mathcal{R}e\,[\mathcal{A} \cdot (z-z')] \; =$$

$$|A^+(\sigma)z|^2 \; - \; |A^-(\sigma)z|^2 \; - \; |A^+(\sigma)z'|^2 \; + \; |A^-(\sigma)z'|^2 \; + \; |P_0(\sigma)(z-z')|^2 \; -$$

$$2\mathcal{R}e\,\left[[{}^t A^+(\sigma)\bar{z}']\cdot[A^+(\sigma)(z-z')] \; - \; [{}^t A^-(\sigma)\bar{z}]\cdot[A^-(\sigma)(z-z')] \right] \; =$$

$$|A^+(\sigma)(z-z')|^2 \; + \; |A^-(\sigma)(z-z')|^2 \; + \; |P_0(\sigma)(z-z')|^2 \; ,$$

whence, for some constant $c_0 > 0$ and all z, $z' \in \mathbf{C}^n$ and all $\sigma \in \mathcal{U} \cap \mathbf{R}^d$:

(III.4.23) $\sigma \cdot [\mathcal{Q}(z) - \mathcal{Q}(z')] - \mathcal{R}e\,[\mathcal{k} \cdot (z-z')] \geq c_0 |z-z'|^2$.

Finally we choose

(III.4.24) $\mathcal{k}_0(z,z',\sigma,t) = \mathcal{k}((1-t)z,z',\sigma)$.

Let us check rapidly that $\mathcal{k}_0$ and, as a consequence, $\mathcal{l}$ and a have all the desired properties. First of all, since

$$| [\mathcal{k}(z,z',\sigma) - \mathcal{k}((1-t)z,z',\sigma)] \cdot (z-z')| \leq Kt|z| \ ,$$

we see that (after a redefinition of c_0 and K) (III.4.20) is a consequence of (III.4.23). It is evident that $\mathcal{k}_0\big|_{t=1}$ and therefore also $a_+\big|_{t=1}$ are independent of (z,s).

Finally, we return to (III.4.22) and we take into account the fact that the dimension of the null-space of $A^-\oplus P_0$ is exactly equal to ν , and therefore the rank of $\overline{\partial}_z \mathcal{k}_0$ is exactly equal to $n - \nu$. This proves the part of the statement (III.4.13) that concerns a_+ .

In dealing with a_- we ought to mention that the requirement that $a_-\big|_{t=1}$ be independent of (z,s) requires that $a_-(z,s,z',\sigma,t) = \tilde{a}_-((1-t)z,(1-t)s,z',\sigma,t)$ for the appropriate choice of the map $\tilde{a}$, which explains the difference between the inequalities (III.4.11) and (III.4.12). ●

We define, for any $\sigma \in \mathbb{R}^d\setminus\{0\}$,

(III.4.25) $a_\pm(z,s,z',s',\sigma,t) = |\sigma| \, a_\pm(z,s,z',s',\dot{\sigma},t) \quad (\dot{\sigma} = \sigma/|\sigma|)$.

Actually a_+ is independent of s and a_- is independent of s' . However, we underline once more the fact that both a_+ and a_- must be independent of (z,s) when $t = 1$. It is clear that $a_\pm$ satisfies (II.5.2) & (II.5.3) with $\Gamma = \mathbb{R}^d\setminus\{0\}$ and with **an appropriate** choice of κ . We may now combine Lemmas III.4.1 & III.4.2:

LEMMA III.4.3.– *Let* $a_{\pm}$ *be defined as in* (III.4.25) *and* $\mu_{\pm}$ *as in* (III.4.8). *Provided* diam $(\mathscr{D} \times \mathscr{B})$ *is sufficiently small, we have, in* $(\mathscr{D} \times \mathscr{B}) \times (\mathscr{D} \times \mathscr{B}) \times (\mathbb{R}^d \backslash \{0\}) \times [0,1]$,

$$(\text{III.4.26}) \quad \mathscr{I}m \left[\mu_{\pm} \dot{\sigma} \cdot [s + \imath\varphi(z,s) - s' - \imath\varphi(z',s')] - \imath a_{\pm}(z,s,z',s',\mu_{\pm}\dot{\sigma},t)(z-z') \right] \geq$$

$$c_0 |z-z'|^2 + \tfrac{1}{2}|s-s'|^2 - C_0(|z|^2 + |s|^2)^{\frac{1}{2}}$$

for suitable positive constants c_0 , C_0 .

Lemma III.4.3 enables us to apply Theorems II.7.1, II.7.2 & II.7.3.

We shall make one of the choices (III.4.25); (III.4.26) entails that conditions (II.6.6) at $t = 0$ and (II.6.11) (with a suitable choice of $\hat{\mathcal{O}}_0$) are satisfied.

First we assume $q > n - \nu$ and take $a = a_+ \big|_{t=0}$ (for all $t \in [0,1]$). By (III.4. 13) condition (II.2.10), hence also (II.5.10), is satisfied. Since a is independent of t property (II.5.11) is trivially satisfied. We may state (*cf.* Theorems III.3.1, III.3.2):

THEOREM **III.4.1.–** *Suppose that* (II.4.1) & (III.4.1) *hold, and that* $n - \nu < q \leq n$. *Then the homotopy formula* (II.7.3) *is valid for all* $f \in C^{\infty}(\mathscr{C} \mathscr{l} \hat{\mathcal{O}}; \Lambda^{m,q})$.

Lastly we look at the "small" values of q . In this case we choose $a = a_-$. Once again, thanks to (III.4.14) we get that (II.2.10) is valid, for all $(z,s,z',s',\sigma) \in \hat{\mathcal{O}} \times \partial\hat{\mathcal{O}} \times (\mathbb{R}^d \backslash \{0\})$ when $t = 1$. From (III.4.13) we derive that (II.5.11) holds provided $q < \nu - 1$, whereas (II.5.10) & (II.5.12) hold provided $q < \nu$.

THEOREM III.4.2.– *Suppose that* (II.4.1) & (III.4.1) *hold. If* $q < \nu - 1$ *the homotopy formula* (II.7.3) *is valid for all* $f \in C^{\infty}(\mathscr{C} \mathscr{l} \hat{\mathcal{O}}; \Lambda^{m,q})$. *If* $q < \nu$ *the right–inverse formula* (II.7.3) *is valid for all* $f \in C^{\infty}(\mathscr{C} \mathscr{l} \hat{\mathcal{O}}; \Lambda^{m,q})$ *such that* $\mathrm{d}f \equiv 0$.

REFERENCES

[Ai–He] Airapetyan, R. A. & Henkin, G. M. – *Integral representations of differential forms on Cauchy–Riemann manifolds and the theory of CR functions*, Uspehi Mat. Nauk, **39**, No 3 (1984), 39–106 ; English transl. in Russian Surveys **39**, No 3 (1984), 41–118.

[A–H] Andreotti, A. & Hill, C. D.– *E. E. Levi convexity and the Hans Lewy problem*, I ; II, Ann. Scuola Norm. Sup. Pisa, **26** (1972), 325–363 and 747–806.

[B–C–T] Baouendi, M. S, Chang, C. H. & Treves, F.– *Microlocal hypo–analicity and extension of CR functions*, J. Diff. Geom. **18** (1983), 331–391.

[B–T] Baouendi, M. S. & Treves, F.– *Unique continuation in CR manifolds and in hypo–analytic structures*, Arkiv för Matematik, **26** (1988), 21–40.

[Be–And] Berndtsson, B. & Andersson, M.– *Henkin–Ramirez formulas with weight factors*, Ann. Institut Fourier **XXXII** (1982), 91–110.

[Bo–Sh] Boggess, A. & Shaw, M. C.– *A kernel approach to the local solvability of the tangential Cauchy–Riemann equations*, Trans. Amer. Math. Soc. **289** (1985), 643–658.

[Bou–Sj] Boutet de Monvel, L. & Sjöstrand, J.– *Sur la singularité des noyaux de Bergman et de Szegö*, Astérisque **34–35** (1976), 123–164.

[Br–Ia] Bros, J. & Iagolnitzer, D.– *Support essentiel et structure analytique des distributions*, Sém. Goulaouic–Lions–Schwartz 1972, No. 18.

[Ha–Po] Harvey, R. & Polking, J.– *Fundamental solutions in complex analysis I, II*, Duke Math. J. **46** (1979), 253–340.

[He] Henkin, G. M.– *H. Lewy's equation and analysis on pseudoconvex manifolds* I (Russian), Uspehi Mat. Nauk. **32** (1977), 57–118; English transl. in Russian Math. Surveys **32** (1977), 59–130; and II, Mat. Sb. **102** (1977), 71–108; Engl. transl. in Math. USSR-Sb. **102** (1977), 63–64.

[He–L] Henkin, G. M. & Leiterer, J.– *Theory of functions on complex manifolds*, Akademic–Verlag Berlin 1983.

[Hö] Hörmander, L.– *Pseudo–differential operators and non–elliptic boundary problems*, Ann. of Math. **83** (1966), 129–209.

[J] Jacobowitz, H.– *On the intersection of varieties with a totally real submanifold*, Proceed. A. M. S. **101** (1987), 127–130.

[Ka–Sch] Kashiwara, M. & Schapira, P.– *A vanishing theorem for a class of systems with simple characteristics*, Invent. Math. **82** (1985), 579–592.

[K–N] Kohn, J. J. & Nirenberg, L.– *A pseudoconvex domain not admitting a holomorphic support function*, Math. Ann. **201** (1973), 265–268.

[K–R] Kohn, J. J. & Rossi, H.– *On the extension of holomorphic functions from the boundary of a complex manifold*, Ann. of Math. **81** (1965), 451–472.

[Lewy] Lewy, H.– *An example of a smooth linear partial differential equation without solution*, Ann. of Math. **66** (1957), 155–158.

[Na–R] Nagel, A. & Rosay, J.–P.– *Non–existence of homotopy formula, for (0,1) forms on hypersurfaces in $\mathbb{C}^3$*, Duke J. **58** (1989), 823–827.

[R] Rosay, J.–P.– *Some applications of Cauchy–Fantappié forms to (local) problems on $\overline{\partial}_b$* , Annali Scuola Norm. Sup. Pisa **XIII** (1986), 225–243.

[Schw] Schwartz, L.– *Théorie des distributions*, 2nd ed., Hermann Paris 1966.

[T] Treves, F.– *Microlocal cohomology in hypo–analytic structures*, in Partial Differential Equations, Springer Lecture Notes No 1324.

Mathematics Department
Rutgers University
New Brunswick, N.J. 08903

MEMOIRS of the American Mathematical Society

SUBMISSION. This journal is designed particularly for long research papers (and groups of cognate papers) in pure and applied mathematics. The papers, in general, are longer than those in the TRANSACTIONS of the American Mathematical Society, with which it shares an editorial committee. Mathematical papers intended for publication in the Memoirs should be addressed to one of the editors:

Ordinary differential equations, partial differential equations and applied mathematics to ROGER D. NUSSBAUM, Department of Mathematics, Rutgers University, New Brunswick, NJ 08903

Harmonic analysis, representation theory and Lie theory to ROBERT J. ZIMMER, Department of Mathematics, University of Chicago, Chicago, IL 60637

Abstract analysis to MASAMICHI TAKESAKI, Department of Mathematics, University of California, Los Angeles, CA 90024

Classical analysis (including complex, real, and harmonic) to EUGENE FABES, Department of Mathematics, University of Minnesota, Minneapolis, MN 55455

Algebra, algebraic geometry and number theory to DAVID J. SALTMAN, Department of Mathematics, University of Texas at Austin, Austin, TX 78713

Geometric topology and general topology to JAMES W. CANNON, Department of Mathematics, Princeton University, Princeton, NJ 08544

Algebraic topology and differential topology to RALPH COHEN, Department of Mathematics, Stanford University, Stanford, CA 94305

Global analysis and differential geometry to JERRY L. KAZDAN, Department of Mathematics, University of Pennsylvania, E1, Philadelphia, PA 19104-6395

Probability and statistics to BURGESS DAVIS, Departments of Mathematics and Statistics, Purdue University, West Lafayette, IN 47907

Combinatorics and number theory to CARL POMERANCE, Department of Mathematics, University of Georgia, Athens, GA 30602

Logic, set theory and general topology to JAMES E. BAUMGARTNER, Department of Mathematics, Dartmouth College, Hanover, NH 03755

Automorphic and modular functions and forms, geometry of numbers, multiplicative theory of numbers, zeta and L-functions of number fields and algebras to AUDREY TERRAS, Department of Mathematics, University of California at San Diego, La Jolla, CA 92093

All other communications to the editors should be addressed to the Managing Editor, RONALD L. GRAHAM, Mathematical Sciences Research Center, AT&T Bell Laboratories, 600 Mountain Avenue, Murray Hill, NJ 07974.

General instructions to authors for

PREPARING REPRODUCTION COPY FOR MEMOIRS

> **For more detailed instructions send for AMS booklet, "A Guide for Authors of Memoirs."**
> **Write to Editorial Offices, American Mathematical Society, P.O. Box 6248,**
> **Providence, R.I. 02940.**

MEMOIRS are printed by photo-offset from camera copy fully prepared by the author. This means that, except for a reduction in size of 20 to 30%, the finished book will look exactly like the copy submitted. Thus the author will want to use a good quality typewriter with a new, medium-inked black ribbon, and submit clean copy on the appropriate model paper.

Model Paper, provided at no cost by the AMS, is paper marked with blue lines that confine the copy to the appropriate size. Author should specify, when ordering, whether typewriter to be used has **PICA**-size (10 characters to the inch) or **ELITE**-size type (12 characters to the inch).

Line Spacing — For best appearance, and economy, a typewriter equipped with a half-space ratchet — 12 notches to the inch — should be used. (This may be purchased and attached at small cost.) Three notches make the desired spacing, which is equivalent to 1-1/2 ordinary single spaces. Where copy has a great many subscripts and superscripts, however, double spacing should be used.

Special Characters may be filled in carefully freehand, using dense black ink, or **INSTANT** ("rub-on") **LETTERING** may be used. AMS has a sheet of several hundred most-used symbols and letters which may be purchased for $5.

Diagrams may be drawn in black ink either directly on the model sheet, or on a separate sheet and pasted with rubber cement into spaces left for them in the text. Ballpoint pen is not acceptable.

Page Headings (Running Heads) should be centered, in CAPITAL LETTERS (preferably), at the top of the page — just above the blue line and touching it.

LEFT-hand, EVEN-numbered pages should be headed with the AUTHOR'S NAME;

RIGHT-hand, ODD-numbered pages should be headed with the TITLE of the paper (in shortened form if necessary).

Exceptions: PAGE 1 and any other page that carries a display title require NO RUNNING HEADS.

Page Numbers should be at the top of the page, on the same line with the running heads.

LEFT-hand, EVEN numbers — flush with left margin;

RIGHT-hand, ODD numbers — flush with right margin.

Exceptions: PAGE 1 and any other page that carries a display title should have page number, centered below the text, on blue line provided.

FRONT MATTER PAGES should be numbered with Roman numerals (lower case), positioned below text in same manner as described above.

MEMOIRS FORMAT

> **It is suggested that the material be arranged in pages as indicated below.**
> **Note: Starred items (*) are requirements of publication.**

Front Matter (first pages in book, preceding main body of text).

Page i — *Title, *Author's name.

Page iii — Table of contents.

Page iv — *Abstract (at least 1 sentence and at most 300 words).

Key words and phrases, if desired. (A list which covers the content of the paper adequately enough to be useful for an information retrieval system.)

*1980 Mathematics Subject Classification (1985 Revision). This classification represents the primary and secondary subjects of the paper, and the scheme can be found in Annual Subject Indexes of MATHEMATICAL REVIEWS beginnning in 1984.

Page 1 — Preface, introduction, or any other matter not belonging in body of text.

Footnotes: *Received by the editor date.
Support information — grants, credits, etc.

First Page Following Introduction – Chapter Title (dropped 1 inch from top line, and centered). Beginning of Text.

Last Page (at bottom) – Author's affiliation.